Animal Cell Substrates for the Manufacture of Biological Products

WHO Documents (1959 – 2012)

Animal Cell Substrates for the Manufacture of Biological Products

WHO Documents (1959 – 2012)

John Petricciani, MD, MS

&

Ivana Knezevic, MD, PhD

Editors

International Alliance for Biological Standardization (IABS)

2017

First Printing: 2017

ISBN: 978-1-387-35716-1

IABS
Rue de la Vallée, 3
CH-1204 Geneva
Switzerland

iabs@iabs.org

Ordering Information: contact the publisher at the above listed address.

Contents

Introduction

Many biological medicinal products are derived from cells that either naturally produce the substance desired, are genetically modified to do so, or as in the case of viral vaccines, act as a substrate in which the vaccine virus replicates to produce large quantities of the vaccine. Traditionally, animal cells have been used for the production of viral vaccines, although considerable progress has been made in the use of non-mammalian cell lines derived from plants and insects.

WHO has regularly reviewed current practices on the use of cells as production substrates, evaluated the risks inherent in specific cell types, and made recommendations to manufacturers and regulatory authorities on their correct handling and use. Over the years, WHO published eight documents on this topic, as described in the following paragraphs.

The first WHO Requirements for cell cultures used in the production of biologicals were formulated in 1959 as part of the production of inactivated poliomyelitis vaccine (IPV) [1]. Specifically, the following two sentences are the only guidance provided regarding the monkey kidney cell substrate. "Virus for the preparation of vaccine shall be grown by aseptic methods in cultures of monkey-kidney cells that have not been propagated in series. The maintenance medium shall contain no protein. If animal serum is used in the propagation of cells, the final vaccine shall not contain more animal serum than one part per million."

Human diploid cells (HDCs) were developed in the 1960s as an alternative to primary cell cultures such as those from monkey kidneys used for polio vaccines. WHO published Requirements for poliomyelitis vaccine (oral) in 1972 that included a section specific for HDCs [2]. Special requirements for HDCs were introduced and include: the concept of a cell seed; a test for tumorigenicity (heterotransplantability); chromosomal monitoring; and an upper limit of acceptability for population doublings.

Advances in technology of production and test methods led to a revision of the Requirements for Poliomyelitis Vaccine (Inactivated) in 1982 [3]. Of particular importance was the recognition that the virus may be grown in well characterized non-tumorigenic continuous cell lines (CCLs). The revised Requirements included a section specific for non-tumorigenic CCLs, as had been done in 1972 for HDCs. The 1982 document introduced the concept of testing for tumorigenicity at 10 passages beyond that used for vaccine production. The amount of DNA in the purified product was recommended to be reduced by a factor of at least 10^8 from that in the initial virus harvest.

With the introduction of monoclonal antibodies and recombinant DNA techniques in the development of vaccines and biotherapeutics in the 1980s, there was a need for additional guidance on the use of CCLs for the production of biological products. WHO convened a study group (SG) in 1986 to consider the issues and to provide recommendations. The report of the SG was published in 1987 [4] and identified viruses as the major element of concern. The upper limit of DNA was recommended to be 100 pg per dose. Transforming proteins were considered to be an insignificant risk factor.

WHO Requirements for CCLs were published in 1987 [5] and took into consideration the SG report. The previous Requirements for CCLs [3] were limited to non-tumorigenic cells. The 1987 document provided general guidance on the characterization of CCLs when used as substrates for the manufacture of biological products, including recombinant DNA products. Specific requirements for purity, including the amount of DNA in a dose, were deferred to individual product Requirements since acceptable levels would depend on many factors including the intended use and the manufacturing process.

The WHO Requirements for cell substrates used for production of biologicals were revised in 1996 and published in 1998 [6]. They describe the characterization and testing of CCLs and diploid cell substrates, along with the general manufacturing requirements applicable to primary cell substrates. Residual cellular DNA was no longer regarded as a significant risk factor that required removal to extremely low levels, and 10 ng of DNA per dose was considered acceptable. In addition, the requirements for diploid cells were relaxed.

The 2003 addendum to the 1996 Requirements was published in 2005 [7], and reclassifies the WHO Vero 10-87 cell line from a master cell bank to a cell seed from which manufacturers can establish a master cell bank through additional qualification.

In 2010, the Expert Committee on Biological Standardization endorsed the proposed WHO recommendations for the evaluation of animal cell cultures as substrates to manufacture biological medicinal products and for the characterization of cell banks [8]. The major changes introduced in that document are:

1. updated general manufacturing recommendations applicable to all types of cell culture production;
2. considerations for the evaluation of new cell substrates such as insect cells and stem cells (SCs);
3. addition of a new section on risk reduction strategies during the manufacture of biological products;
4. addition of a section on Good Cell Culture Practice;
5. addition of detailed methods used to test for bovine viruses in serum;
6. updated section on tumourigenicity testing, with the addition of a model protocol for the nude mouse;
7. addition of a section on oncogenicity testing of tumourigenic cell lysates, and a model protocol;
8. no recommendations for acceptable levels of residual cellular DNA since they are product specific

This volume has collected together these eight WHO documents to serve as an historical reference and to facilitate an understanding of the evolution of issues and positions that have been taken since the 1950s.

We are grateful to WHO for granting permission to reproduce these documents which are referenced below.

John Petricciani, MD, MS
IABS

Ivana Knezevic, MD, PhD
WHO

References

1. Requirements for poliomyelitis vaccine (inactivated). In: Requirements for biological substances: 2. Requirements for poliomyelitis vaccine (inactivated). Report of a Study Group. Geneva. World Health Organization, 1959 (WHO Technical Report Series, No. 178).
2. Requirements for poliomyelitis vaccine (oral). (Requirements for biological substances, No.7). Annex 1, Part C in: WHO Expert Committee on Biological Standardization. Twenty-fourth report. Geneva. World Health Organization, 1972 (WHO Technical Report Series, No. 486).
3. Requirements for poliomyelitis vaccine (Inactivated). (Requirements for biological substances, No.2). Annex 2, Part D in: WHO Expert Committee on Biological Standardization. Thirty-second report. Geneva. World Health Organization, 1982 (WHO Technical Report Series, No. 673).
4. Report of a WHO Study Group on Biologicals, Acceptability of cell substrates for production of biologicals. Geneva. World Health Organization, 1987 (WHO Technical Report Series, No. 747).
5. Requirements for continuous cell lines used for biologicals production. Annex 3 in: WHO Expert Committee on Biological Standardization. Thirty-sixth report. Geneva. World Health Organization, 1987 (WHO Technical Report Series, No. 745).
6. Requirements for the use of animal cells as in vitro substrates for the production of biologicals. Annex 1 in: WHO Expert Committee on Biological Standardization. Forty-seventh report. Geneva. World Health Organization, 1998 (WHO Technical Report Series, No. 878).
7. Requirements for the use of animal cells as in vitro substrates for the production of biologicals (Addendum 2003). Annex 4 in: WHO Expert Committee on Biological Standardization. Fifty-fourth report. Geneva. World Health Organization, 2005 (WHO Technical Report Series, No. 927).
8. Recommendations for the evaluation of animal cell cultures as substrates for the manufacture of biological medicinal products and for the characterization of cell banks. Replacement of TRS 878, Annex 1. Annex 3 in WHO Expert Committee on Biological Standardization. Sixty-first report. Geneva. World Health Organization, 2013 (WHO Technical Report Series, No. 978).

This report contains the collective views of an international group of experts and does not necessarily represent the decisions or the stated policy of the World Health Organization.

REQUIREMENTS FOR BIOLOGICAL SUBSTANCES

1. General Requirements for Manufacturing Establishments and Control Laboratories
2. Requirements for Poliomyelitis Vaccine (Inactivated)

Report of a Study Group

World Health Organization
Technical Report Series (TRS)
178

World Health Organization, Palais des Nations, Geneva **1959**

STUDY GROUP ON GENERAL REQUIREMENTS FOR MANUFACTURING ESTABLISHMENTS AND CONTROL LABORATORIES AND ON REQUIREMENTS FOR POLIOMYELITIS VACCINE

Geneva, 2-7 June 1958

Members:

Dr O. Bonin, Scientific Member, Institute for Chemotherapeutic Research, Paul-Ehrlich-Institut, Frankfurt-am-Main, Federal Republic of Germany

Dr D. G. Evans, Director, Biological Standards Control Laboratories, Medical Research Council, Hampstead, London, England

Dr S. Gard, Professor of Virus Research, School of Medicine, Karolinska Institutet, Stockholm, Sweden (*Rapporteur*)

Dr J. H. S. Gear, Director of Research, Poliomyelitis Research Foundation Laboratories, Johannesburg, Union of South Africa (*Chairman*)

Dr A. Lafontaine, Director, Institut d'Hygiène et d'Epidémiologie, Brussels, Belgium

Dr P. Lépine, Chef du Service des Virus, Institut Pasteur, Paris, France (*Vice-Chairman*)

Dr H. von Magnus, Chief, Department of Poliovirus, Statens Seruminstitut, Copenhagen, Denmark

Dr R. Murray, Director, Division of Biologics Standards, National Institutes of Health, Bethesda, Md., USA

Dr G. Penso, Chief, Laboratory of Microbiology, Istituto Superiore di Sanità, Rome, Italy

Dr V. Soloviev, Scientific Director, Moscow Institute for Poliomyelitis Prophylactics, Moscow, USSR.

Secretariat:

Dr B. K. Bhattacharya, Medical Officer, Biological Standardization, WHO

Dr N. K. Jerne, Chief Medical Officer, Biological Standardization, WHO (*Secretary*)

Dr U. Krech, Chief, Virus Department, Serum and Vaccine Institute, Berne, Switzerland (*Consultant*)

Dr A. M.-M. Payne, Chief Medical Officer, Endemo-Epidemic Diseases, WHO

This report was originally issued as mimeographed document WHO/BS/IR/44.

REQUIREMENTS FOR BIOLOGICAL SUBSTANCES
Report of a Study Group

1. GENERAL CONSIDERATIONS

The Study Group further considered a proposed first draft of requirements for poliomyelitis vaccine,[2] the recommendations of the second report of the Expert Committee on Poliomyelitis,[3] a document on sterility control,[4] several documents concerning potency control,[5] as well as a large amount of unpublished data that had been collected by the members of the Group.

The Study Group surveyed the regulations and requirements for the manufacture and control of poliomyelitis vaccine that had been adopted in some countries. Most of these had been modeled on the requirements originally formulated in the United States of America. However, a survey of the documents submitted to the Study Group[6] and of the additional information presented by members of the Group showed that there were many differences between the requirements now in use in individual countries. The Group agreed that an important purpose would be served by the formulation of essential requirements for the manufacture and control of poliomyelitis vaccine which would be internationally acceptable.

The Study Group drafted the requirements for poliomyelitis vaccine given in Annex 2 which it considered to be generally applicable to all vaccines in current use and which are therefore based on inactivated, trivalent vaccines only. The Group agreed that situations could arise in which a largely non-immune population might be threatened by an epidemic of one particular poliovirus type, and that in such situations the use of monovalent, or even divalent, vaccines might appear desirable.

---

2. WHO Secretariat, unpublished working documents WHO/BS/IR/29 & WHO/BS/IR/30
3. *Wld Hlth Org. techn. Rep. Ser.,* 1958, **145**
4. Eissner, G. & Bonin, O., unpublished working document WHO/BS/IR/36
5. Gard, S., Johnsson, T., Lycke, E., Melén, B., Olin, G., Salenstadt, R.-& Wrange, G., unpublished working document WHO/BS/IR/32; de Somer, P., unpublished working document WHO/BS/IR/33; Lépine, P., Roger, F. & Sautter, V., unpublished working document WHO/BS/IR/34; Krech, U., unpublished working document WHO/BS/IR/38; Benyesh, M. & Melnick, J. L., unpublished working document WHO/BS/IR/39; United States of America, Division of Biologics Standards, National Institutes of Health, unpublished working document WHO/BS/IR/42; Prigge, R., Günther, O. & Bonin, O., unpublished working document WHO/BS/IR/43
6. Lafontaine, A., unpublished working document WHO/BS/IR/40 (Belgian requirements) ; Canada, Laboratory of Hygiene, unpublished working document WHO/BS/IR/41; von Magnus, H., unpublished working document WHO/BS/IR/35 (Danish requirements); Lépine, P., unpublished working document WHO/Polio/23 (French requirements); Federal Republic of Germany, Paul-Ehrlich-Institut, Frankfurt-am-Main, unpublished third draft of the provisional regulations for the national control of poliomyelitis vaccines; Penso, G., unpublished working -document WHO/BS/IR/37 (Italian requirements) ; England and Wales, *Therapeutic Substances Amendment Regulations, 1956,* Schedule, Part XVII; United States of America, *Code of Federal Regulations, 1956,* Title 42, Chapter I, Part 73.

3. REVISION OF TEXT OF THE REQUIREMENTS FOR POLIOMYELITIS VACCINE

The Study Group discussed what procedures could be recommended for revisions of the text of the Requirements for Poliomyelitis Vaccine now drawn up (Annex 2 to this report), and it recommended that the Secretariat of the World Health Organization submit all comments that might be forthcoming from members of the Group, or from other experts, to appropriate study groups or expert committees for consideration, in order to arrive at definitive requirements.

At appropriate intervals, the requirements should be re-examined and provision should be made for the issue of a revised text whenever this became necessary.

Annex 2

REQUIREMENTS FOR POLIOMYELITIS VACCINE (INACTIVATED) (REQUIREMENTS FOR BIOLOGICAL SUBSTANCES No. 2)

General Considerations

The recommendation of international requirements for inactivated poliomyelitis vaccine is complicated by the fact that a number of different manufacturing and testing procedures are in use in various countries. The procedures differ mainly in the incorporation of different virus strains, the inactivation and filtration methods, and the use of preservatives and adjuvants in the final vaccine. In spite of these differences it is felt that certain essential requirements concerning manufacture and control can be formulated. The present recommendations are based on methods currently in use and future revisions will be necessary.

Each of the following sections constitutes a recommendation. The parts of each section which are printed in large type have been written in the form of requirements so that, if a health administration so desires, these parts as they appear may be used as definitive national requirements. The parts of each section which are printed in small type concern points on which comments seemed desirable.

3.1 *Control of source materials*

3.1.1 *Virus strains*

Strains of poliovirus used in the production of vaccine shall be identified by historical records, infectivity tests, and by immunological methods. Any strain which will yield a vaccine meeting the requirements set forth in the present document may be used. Production of vaccine shall be based on a seed virus system; the poliovirus used for vaccine production shall not have passed more than ten subcultures, counted from a strain culture on which the original laboratory and field tests were done.

Preference should be given to strains of low pathogenicity to monkeys.
Each new seed lot of a strain of low pathogenicity should be retested for virulence before being used as a seed virus. Samples of the strains used should be deposited in the national control laboratory. Strains of poliovirus currently used in production of vaccine may be obtained by application direct or through the World Health Organization to specialized laboratories.*

* The following laboratories have expressed their willingness to supply samples of strains for this purpose: Institut d'Hygiène et d'Epidémiologie, Brussels, Belgium; Statens Seruminstitut, Copenhagen, Denmark; Institut Pasteur, France; Paul-Ehrlich-Institut, Frankfurt-am-Main, Federal Republic of Germany; Laboratory of Microbiology, Istituto Superiore di Sanità, Rome, Italy; The Poliomyelitis Research Foundation Laboratories, Johannesburg, South Africa; Statens Bacteriologiska Laboratorium, Stockholm, Sweden;

Biological Standards Control Laboratories, Medical Research Council, London, N.W.3, England; Division of Biologics Standards, National Institutes of Health, Bethesda, Md., USA; Moscow Institute for Poliomyelitis Prophylactics, Moscow, USSR.

3.1.2 *Monkeys*

Suitable species of monkeys, in good health, shall be used as the source of kidney tissue for the production of poliovirus. Each animal shall be examined at necropsy for signs of disease and, if there is any pathological lesion of significance with regard to their use in the preparation of the vaccine, the kidneys shall be discarded. Kidney tissue from monkeys that have been used previously for experimental purposes shall not be used. An exception can be made in the case of monkeys used for the safety or potency tests with negative clinical findings.

It is recommended that monkeys be kept in as small groups as possible in order to reduce dissemination of infections within the colony.

3.1.3 *Tissue culture for virus production*

Virus for the preparation of vaccine shall be grown by aseptic methods in cultures of monkey-kidney cells that have not been propagated in series. The maintenance medium shall contain no protein. If animal serum is used in the propagation of cells, the final vaccine shall not contain more animal serum than one part per million.

Suitable antibiotics in minimum concentrations required for sterility may be used. If penicillin is used its concentrations may not exceed 200 International Units per ml. Non-toxic pH indicators may be added, e.g., phenol red in a concentration of 0.002%.

This report contains the collective views of an international group of experts and does not necessarily represent the actions or the stated policy of the World Health Organization

WHO Expert Committee on Biological Standardization

Twenty-fourth Report

World Health Organization
Technical Report Series (TRS)
486

World Health Organization, Geneva **1972**

WHO EXPERT COMMITTEE ON BIOLOGICAL STANDARDIZATION

Geneva, 3-9 November 1971

*Members:**

Dr I. Archetti, Chief, Virus Department, Istituto Superiore di Sanità, Rome, Italy
Dr H. H. Cohen, Director, Rijks Instituut voor de Volksgezondhcid, Bilthoven, Netherlands
Dr J. Desbordes, Director, Microbiology Section, Laboratoire des Actions de Santé, Paris, France (*Vice-Chairman*)
Dr N. K. Dutta, Director, Haffkine Institute, Bombay, India
Dr L. Higy-Mandic, Chief, Department of Biological Standardization, Institute of Immunology, Zagreb, Yugoslavia
Dr D. W. Howes, Chief Virologist, Viral Products Section, National Biological Standards Laboratory, Parkville, Victoria, Australia (*Rapporteur*)
Mr J. W. Lightbown, Division of Biological Standards, National Institute for Medical Research, London, England
Dr H. Mirchamsy, Associate Director, Razi State Institute, Teheran, Iran
Dr R. Murray, Director, Division of Biologics Standards, National Institutes of Health, Bethesda, Md., USA (*Chairman*)
Dr F. P. Nagler, Chief, Virus Laboratories, Department of National Health and Welfare, Ottawa, Ont., Canada
Dr J. Spaun, Deputy Director, Department of Biological Standardization, Statens Seruminstitut, Copenhagen, Denmark

Secretariat:

Dr D. R. Bangham, Director, Division of Biological Standards, National Institute for Medical Research, London, England (*Consultant*)
Dr P. Krag, Director, Department of Biological Standardization, Statens Seruminstitut, Copenhagen, Denmark (*Consultant*)
Dr A. S. Outschoorn, Chief Medical Officer, Biological Standardization, WHO, Geneva, Switzerland (*Secretary*)
Dr W. W. Wright, Director, Division of Drug Biology, Food and Drug Administration, Department of Health, Education and Welfare, Washington, D.C., USA (*Consultant*)

**Unable to attend:*

Dr S. G. Dzagurov, Director, Tarasevic State Institute for the Control of Medical Biological Preparations, Moscow, USSR
Professor D. G. Evans, Director, Lister Institute of Preventive Medicine, London, England
Professor H. O. Schild, Pharmacology Department, University College, London, England

INTERNATIONAL REQUIREMENTS FOR BIOLOGICAL SUBSTANCES

30. Requirements for Poliomyelitis Vaccine (Oral)

The Committee studied the revised requirements for poliomyelitis vaccine (oral) that had been prepared by the WHO Secretariat[4] in collaboration with a number of experts. The original requirements were published in 1962 and the first revision in 1965. After making certain modifications, the Committee agreed that the present text of these requirements, which now included requirements for vaccine prepared using monkey kidney cells or human diploid cells, was satisfactory and that they would be useful for the control of poliomyelitis vaccine (oral) produced in different countries.

The Committee adopted these requirements and agreed that they should be annexed to the present report (see Annex 1).

---

[4] Unpublished working document WHO/BS/71.1027 Rev.1.

Part C. Requirements for Poliomyelitis Vaccine (Oral) Prepared Using Human Diploid Cells

The following additional or alternative requirements are for poliomyelitis vaccine (oral) prepared with human diploid cells and concern the testing of the cell substrate used for production of the vaccine; they should therefore be added to or substituted for the appropriate sections in Parts A and B as indicated. **All the other requirements given in Parts A and B of the document are applicable to this vaccine as well.**

Modifications affecting Part A. Manufacturing Requirements

1. Definitions

1.4 *Terminology*

The following shall apply instead of the definition given in Part A, section 1.4:

Single harvest: A virus suspension of one virus type harvested from cell cultures prepared from one ampoule of cell seed.

The following shall be added:

Cell seed: A quantity of cells derived from a single human tissue and of uniform composition, stored frozen at -70°C or below in aliquots, one of which would be used for the production of each single harvest.

2. General manufacturing requirements

The following shall be added:

No cell cultures other than those approved by the national control authority for the production of this vaccine shall be introduced or handled in the production area.

3. Production control

3.1 *Control of source materials*

The following shall apply instead of the requirements given in Part A, section 3.1.2 :

3.1.2 *Cell seed for the production of cell cultures*

The production of the human diploid cell cultures used for vaccine manufacture shall be based on the cell seed system.

The cell seed used for the production of oral poliomyelitis vaccine shall be that approved by and registered with the national control authority. The cells shall have been characterized with respect to their genealogy, growth characteristics, storage conditions, and karyology, and they shall have been shown, by tests in animals and eggs, to be free from adventitious agents.[1] The supernatant fluids shall also have been shown by tests in cell cultures to be free from adventitious agents. The cells shall be shown to be diploid and stable with respect to karyology and morphology and to meet the requirements given in the subsections of Part C, section 3.2.3 throughout their finite life span. The cells shall at no time have shown the properties of a continuous cell line.

The tests in animals and eggs for adventitious agents shall include one of inoculating cells by the intramuscular route into each of the following groups of animals using at least 10^7 cells for each group:

> 2 litters of suckling mice, totaling at least 10 animals, less than 24 hours old;
>
> 10 adult mice;
>
> 5 guinea-pigs; and
>
> 5 rabbits

as well as into the allantoic cavity of 10 embryonated eggs 9-11 days old.

The animals shall be observed for at least 4 weeks and the embryonated eggs shall be examined after not less than 3 days. Any animals that are sick or show any abnormality shall be investigated to establish the cause of illness. The allantoic fluids shall be tested with guinea-pig and chick or other avian red cells for the presence of haemagglutinins.

The cells shall also be shown to be free from potential heterotransplantability by appropriate tests in animals.

> Suitable tests using immunosuppressed animals may be made as follows: A quantity of 10^6 cells obtained from cultures at the same passage level as those used for vaccine production are inoculated into each cheek pouch of 6 hamsters, 3 of which have been treated with cortisone, and the animals observed for not less than 4 weeks. Some hamsters of the strain used should have been inoculated with HeLa or KB cells and it should have been shown that tumour formation can be caused by the inoculation of the neoplastic tissue, thus demonstrating the ability of the strain of animals to give rise to tumours, Any other test using animals treated with immunosuppressive agents and with equal sensitivity to neoplastic cells may be used.

The cells are suitable for vaccine production if at least 80% of the animals inoculated with the cells remain healthy and survive the observation period, none of the animals or eggs show evidence of the presence in the cell cultures of any adventitious agent, and none of the animals show evidence of tumour formation from the cells.

The cell seed shall also have been shown to yield cell cultures capable of producing vaccine that has been found to be safe and antigenic,

3.1.3 *Virus seed lot system*

The requirements given in this section shall be amended to allow the preparation of seed lots in either human diploid cell cultures or monkey kidney-cell cultures. If human diploid cells are used the requirements given in Part C, sections 3.1.4 and 3.2 shall apply.

3.1.4 *Tests on virus seed lots*

The requirements given in this section shall apply with the exception that the requirements referred to in Part A, section 3.3 shall be applicable as amended in Part C, section 3.3.

3.2 *Production precautions*

The following shall apply instead of the requirements given in Part A, section 3.2, with the exception of section 3.2.1, which is not applicable.

3.2.2 *Cell cultures used for vaccine production*

Only human diploid cell cultures derived from a cell seed approved by the national control authority shall be used for vaccine production. The production of each single harvest shall be initiated from a new ampoule of the cell seed. All processing of the cell seed and subsequent cell cultures shall be done in an area in which no other cells are handled. The cell cultures shall be used only if no changes have occurred in their growth characteristics (including freedom from potential heterotransplantability as shown by the test in Part C, section 3.1.2) and storage conditions and if no changes from the normal karyology have been shown to occur within the total number of population doublings[2] that correspond to the average finite life of the cells as determined under the particular conditions of the production establishment.

> It is important that the karyological pattern should have been determined not to differ from that established for the cell seed for the total number of population doublings of the cell cultures and that no changes from the normal karyology should have been shown to occur within a proportion of the total number of population doublings that would be used for vaccine production.

The cells shall, however, not be used beyond two-thirds of the total number of population doublings corresponding to the average finite life of the cells.

3.2.3 *Tests of cell cultures used for vaccine production*

Each batch of diploid cells used as a substrate for the production of a single harvest of virus shall be tested for freedom from extraneous agents and for retention of normal morphology and karyology. Only cells that have the normal morphological appearance and karyology and that pass the test described below shall be used for vaccine production. The cells shall have been shown to have biological properties unchanged from those of the cell seed as shown by the tests outlined in Part C, section 3.1.2.

In some countries immunobiological tests are made to verify identity of the cells with the cell seed and the purity of the cultures.

Cells sufficient for chromosome monitoring (part C, section 3.2.3.1) and for preparing control cultures (part C, section 3.2.4) shall be taken from the pooled material removed from each culture vessel not earlier than two population doublings preceding the doubling level at which cells are to be inoculated with vaccine virus. These cells, or cells subcultured from these cells, shall be used for making preparations for chromosome monitoring. The remaining cells shall be set aside as control material. The supernatant fluids from the production bottles at the cell doubling level at which they are to be inoculated with vaccine virus shall be used for the tests for extraneous agents and for bacteria, fungi, and mycoplasma (part C, sections 3.2.3.2 and 3.2.3.3).

On the day of inoculation with seed lot virus, each cell culture shall be examined for degeneration caused by an infective agent. If this examination or any of the tests required in this section shows evidence of the presence in a cell culture of any adventitious agent, the poliovirus grown in the whole group of cultures concerned shall not be used for vaccine production.

3.2.3.1 *Chromosome monitoring pool --- preparation and testing*

Preparations shall be made from at least 1% of the sample of pooled cell substrate removed from the culture vessels. Chromosome monitoring shall be done at the stage equivalent to the doubling level at which cells are to be inoculated with vaccine virus, or within three population doublings beyond this stage, but if the cells are subcultivated they shall be repooled. For determination of the general character of the cell material, a minimum of 300 cells shall be examined for frequency of polyploidy and a minimum of 100 metaphase plates for exact counts, and analyses of karyotype shall be performed on at least one selected cell. The metaphase plates shall be examined for characteristics that shall include frequency of chromosome breaks, structural chromosome abnormalities, other abnormalities such as despiralization or marked attenuations of the primary or secondary constrictions, and the presence of minute chromosomes.

For vaccine production examination of the cells is usually made between the 27th and 33rd population doubling and the frequency of cells in metaphase with chromosome breaks should not exceed 9%, with structural abnormalities not more than 4%, and with polyploidy not more than 5%[3]. All cells showing: abnormalities should be subjected to detailed examination and records should be maintained of the detailed criteria applied to particular abnormalities evaluated in the karyotype analysis.

Permanent stained slide preparations of the chromosome monitoring pool, or photographs of these, shall be maintained as part of the record of the batch of vaccine and for monitoring successive batches.

It is desirable that a portion of the sample of pooled cell substrate removed from the culture vessels be stored frozen so as to retain viability. This would be available for future reference for karyology or for any other purpose relating to the batch of vaccine.

3.2.3.2 *Tests of supernatant fluids for extraneous viruses*

The pooled supernatant fluids from the production cell cultures taken immediately prior to inoculation of virus for vaccine production shall be tested by the inoculation of 100ml quantities into cultures from each of the following:

human embryonic kidney cells,

Cercopithecus or *Erythrocebus* monkey kidney cells, and

primary rabbit kidney cells.

In some countries the national control authority requires that additional tests be made in the same cell strain used for vaccine production of another human diploid cell strain and in rhesus monkey kidney cells, as well as the use of larger volumes of pooled supernatant fluid.

Serum used in the nutrient medium of the cultures other than the monkey kidney-cell cultures shall have been shown to be free from inhibitors to virus growth. Each sample shall be inoculated into bottles of the cell cultures, in such a way that the dilution of the pooled fluid in the nutrient medium does not exceed 1 in 4. The area of the cell sheet shall be at least 3 cm^2 per ml of pooled fluid. At least one bottle of the cell cultures shall remain uninoculated and serve as a control.

In the case of the Cercopithecus or Erythrocebus monkey kidney-cell cultures, added serum may be used in the propagation of the cells, provided it does not contain inhibitors to virus growth, but the maintenance medium after inoculation of test material should contain no added serum.

The cultures shall be examined for normal morphology during a period of incubation at 37°C for 14 to 21 days. At the end of the observation period a subculture shall be made from the cercopithecus or erythrocebus cultures in the same cell system and these cultures shall be examined for a further 14 days. Some of the cultures from each monkey kidney shall be tested also for haemadsorbing viruses using appropriate red cells that have not been stored for more than 7 days.

> These tests are usually made using guinea-pig red cells. In some countries the national control authority requires that the tests for haemadsorbing viruses be made in addition using red cells from humans (blood group IV O), monkeys and chickens. The cultures should be examined at 3-5 days and again at 12 days. All tests should be read after incubation for not less than 30 minutes at 0°-4°C and again after a further incubation for at least 30 minutes at 20°-25°C. In addition the test with monkey cells should be read after a still further incubation for 30 minutes at 34°-37°C.

If the samples of supernatant fluid are not tested immediately they shall be kept at a temperature below -60°C.

3.2.3.3 *Tests for bacteria, fungi, and mycoplasma*

A volume of 20 ml of the pooled supernatant fluids from the production cell cultures shall be tested for bacteria and mycotic sterility and for mycoplasma. The tests for bacterial and mycotic sterility shall be made as described in Part A, section 5, of the Requirements for Biological Substances No. 6 (General Requirements for the Sterility of Biological Substances),[4] and the tests for mycoplasma shall be done using both solid and liquid media that have been shown to be capable of supporting the growth of sterol-requiring and non-sterol-requiring mycoplasmas, using 10 ml for each group of tests,

> In some countries the volume of pooled supernatant fluids is ultracentrifuged and both the pellet and its supernatant fluid tested for sterility.

3.2.4 *Tests of control cell cultures*

The cells set aside as control material (see Part A, section 3.2.3) shall be treated in a similar manner to the production cell cultures but remain uninoculated as control cultures for the detection of extraneous viruses.

> The sample of pooled material taken should be such that the area of cells at the stage equivalent to the doubling level at which cells are to be inoculated with vaccine virus, or beyond, represents not less than 25% of that of the total of cell suspension derived from the ampoule of cell seed at that stage.

These control cell cultures shall be incubated under the same conditions as the inoculated cultures for at least 2 weeks, and shall be examined during this period for evidence of cytopathic changes. For the test to be valid, not more than 20% of the control cell cultures may be discarded for non-specific, accidental reasons.

At the end of the observation period, the control cell cultures shall be examined for degeneration caused by an infective agent. If this examination or any of the tests required in this section shows evidence of the presence in a control culture of any adventitious agent, the poliovirus grown in the corresponding inoculated cultures derived from the same ampoule of cell seed shall not be used for vaccine production.

3.2.4.1 *Tests for haemadsorbing viruses*

At the time of inoculating the production cultures with vaccine virus or at the time of virus harvest, cells comprising 4% of the control cells shall be tested for the presence of haemadsorbing viruses using guinea-pig red cells that have not been stored for more than 7 days.

> These tests are usually made using guinea-pig red cells. In some countries the national control authority requires that the tests for haemadsorbing viruses be made in addition using other types of red cells including human (blood group IV O), monkey and chicken or other avian. The cultures should be examined at 3-5 days and again at 12 days. All tests should be read after incubation for 30 minutes at 0-4°C and again after a further incubation for 30 minutes at 20°-25°C. In addition the test with monkey cells should be read after a still further incubation for 30 minutes at 34°-37°C.

3.2.4.2 *Tests for other extraneous agents*

At the time of harvest, or not more than 7 days after the day of inoculation of the production cultures with seed lot virus, samples of 10 ml of the pooled supernatant fluids from the control cultures shall be taken and tested for extraneous viruses in each of the cell cultures, as described in Part C, section 3.2.3.2. At the end of the observation period for the original control cell cultures similar samples of the pooled fluid shall be taken and the tests referred to in this section shall be repeated.

3.2.5 *Temperature of incubation*

After inoculation of the production bottles with virus both inoculated and control cell cultures shall at no time be at a temperature outside the range of 33°-35°C for the relevant periods of incubation.

> The temperature should not vary by more than ± 0.5°C.

3.3 *Control of single harvests*

The following modifications shall apply to the sections indicated below:

3.3.1 *Single harvest*

The following shall be added :

> The inoculated cell cultures should be processed in such a manner that each virus suspension harvested remains identifiable as a single harvest and is kept separate from

other harvests until all the results of tests prescribed for supernatant fluids in Part C, section 3.2.3.2, have been obtained.

3.3.4 *Tests of neutralized single harvests on monkey kidney and human cell cultures*

For the requirements prescribed in this section of Part A the volume of each single harvest taken for neutralization and testing shall be at least 10 ml and shall be such that at least a total of 50 ml or the equivalent of 500 doses of final vaccine, whichever is the greater, has been withheld from the corresponding bulk suspension.

The antisera used for neutralization shall be of non-human origin and shall have been prepared in animals other than monkeys, using antigens cultured in non-simian cells.

The neutralized suspension shall be divided into two portions and tested in *Cercopithecus* or *Erythrocebus* monkey kidney-cell cultures and human cell cultures sensitive to measles, by culturing and subculturing as described in this section of Part A.

Serum used in the nutrient medium of these cultures shall have been shown to be free from inhibitors of virus growth.

3.3.5 *Tests in rabbit kidney-cell cultures for herpes virus and other viruses*

For the requirements prescribed in this section of Part A the volume of each single harvest taken for testing shall be at least 5 ml.[5] Serum used in the nutrient medium of these cultures shall have been shown to be free from inhibitors of virus growth, including herpes virus.

8. Labelling

The relevant requirements regarding the leaflet accompanying the package shall be amended to include statements that the vaccine has been prepared using human diploid cells and that the manufacture of the vaccine has been undertaken with the approval of the national control authority (see Part B below).

Modifications affecting Part B. National Control Requirements

1. General

The following requirements shall be added :

The manufacture of oral poliomyelitis vaccine using human diploid cell cultures shall be undertaken only with the approval of the national control authority, and then only under a system of chromosome monitoring that will ensure that the cell cultures used for vaccine production have not undergone changes that may adversely affect the safety and efficacy of the product. The national control authority shall therefore give directions to the manufacturing establishment concerning the suitability of the cell seed to be used for vaccine production and the acceptable cell population-doubling level that must not be exceeded for the cell cultures derived therefrom; such directions shall also include criteria for acceptable karyology. Since the information necessary to determine the acceptability of a particular cell strain can be gained only from detailed study in a number of laboratories over several years, the national control authority shall take into consideration all available information in making a decision.

It is desirable that a sample of the cell seed used for vaccine production consisting of not less than 10^7 cells should be lodged with the national control authority. The sample should be frozen so as to retain viability and stored at -70°C or below.

Footnotes

1. Other types of test have been suggested, e.g., examination of ultrathin sections under the electron microscope. Information is insufficient, however, to include such tests in these requirements.
2. *Population doubling*. This may be calculated from an actual cell count of an aliquot or by estimation of the area of expansion of the cell sheet.
3. These upper limits correspond to the 99 percentile of the values established from a large number of observations on the cell seed system derived from the human diploid cell strain WI-38. However, greater experience of, and more observations in a number of laboratories on, a particular cell seed may result in revised figures being established for such a percentile. National control authorities should consider all available information in specifying the criteria to be fulfilled (see the additional Requirements to Part B, National Control Requirements). Detailed information is available in the Minutes of Meetings of the Cell Culture Committee of the Permanent Section on Microbiological Standardization (J.A.M.S.) and the Workshop on Karyology of Human Diploid Cells, Chatham, Mass., USA, 1971. These values will not necessarily be applicable if another human diploid cell strain is used.
4. *Wld Hlth Org. techn. Rep. Ser.,* 1960, No. 200, p. 13.
5. The tests should be made as for tests for B virus in rabbit kidney-cell cultures, since B virus may also be a possible, though improbable, contaminant.

This report contains the collective views of an international group of experts and does not necessarily represent the actions or the stated policy of the World Health Organization

WHO Expert Committee on Biological Standardization

Thirty-second Report

World Health Organization
Technical Report Series (TRS)
673

World Health Organization, Geneva **1982**

WHO EXPERT COMMITTEE ON BIOLOGICAL STANDARDIZATION

Geneva, 22-28 September 1981

*Members**

Professor S. G. Dzagurov, Director, Tarasevic State Institute for the Standardization and Control of Medical Biological Preparations, Moscow, USSR
Dr J. W. Lightbown, Head, Division of Antibiotics, National Institute for Biological Standards and Control, London, England (*Chairman*)
Me J. Lyng, Vaccine Department, State Serum Institute, Copenhagen, Denmark
Dr R. Murata, formerly Director-General, National Institute of Health, Tokyo, Japan (*Vice-Chairman*)
Dr R. Netter, Director-General, National Health Laboratory, Paris, France
Dr J. Robbins, Director, Division of Bacterial Products, Bureau of Biologics, Food and Drug Administration, Bethesda, MD, USA (*Rapporteur*)
Me J. R. Thayer, Head, Antibiotics Section, National Biological Standards Laboratory, Canberra, Australia
Dr Tsou Pang-Chu, Chief, Division of Bacterial Vaccines, National Institute for the Control of Pharmaceutical and Biological Products, Temple of Heaven, Beijing, China
Dr W. W. Wright, Senior Scientist, Drug Standards Division, the United States Pharmacopeia and the National Formulary, Rockville, MD, USA

Secretariat

Dr J. N. Ashworth, Vice-President, Scientific Affairs, Biological Products, Cutter Laboratories Inc., Berkeley, CA, USA (*Consultant*)
Dr D.R. Bangham, Head, Division of Hormones, National Institute for Biological Standards and Control, London, England (*Consultant*)
Mr V.F. Davey, Technical Director, Commonwealth Serum Laboratories, Parkville, Victoria, Australia (*Consultant*)
Mr W. Duimel, Central Laboratory of the Netherlands Red Cross Blood Transfusion Service, Amsterdam, Netherlands (*Consultant*)
Professor W. Hennessen, Bureau of Applied Immunology, Berne, Switzerland (*Consultant*)
Dr H. W. Krijnen, Director, Central Laboratory of the Netherlands Red Cross Blood Transfusion Service, Amsterdam, The Netherlands (*Consultant*)
Dr F. H. Meskal, Director, Central Laboratory and Research Institute, Addis Ababa, Ethiopia (*Consultant*)
Dr F. T. Perkins, Chief, Biologicals, WHO, Geneva, Switzerland (*Secretary*)
Dr J. D. van Ramhorst, Biologicals, WHO, Geneva, Switzerland
Dr D. P. Thomas, Head, Division of Blood Products, National Institute for Biological Standards and Control, London, England (*Consultant*)

---

*Unable to attend: Dr Chou Hai-chun, Vice-Director, National Institute for the Control of Pharmaceutical and Biological Products, Beijing, China; Professor B. Lunenfeld, Director, Institute of Endocrinology and Chief of Division of Laboratories, The Chaim Sheba Medical Center, Tel-Hashomer, Israel; Professor F. Oberdoerster, Director, State Institute of Serum and Vaccine Testing of the German Democratic Republic, Berlin.

38. Requirements for Poliomyelitis Vaccine (Inactivated)

The Committee was informed that the advances in technology of production and test methods warranted a complete revision of the Requirements for Poliomyelitis Vaccine (Inactivated). Of particular importance was the recognition that the virus may be grown in well characterized nontumorigenic continuous cell lines, provided that the vaccine is purified and shown to contain no detectable cellular DNA and that tests for freedom from tumorigenicity of the cell lines have been included. Another important advance is the recognition of the need to ensure that the vaccine is sufficiently concentrated and is shown by both *in vitro* and *in vivo* methods to contain an adequate quantity of all three antigenic types.

The Committee was also informed that the test for the detection of virulent virus particles that may have escaped inactivation has been reviewed. The experience over the last 25 years has shown that the inoculation of cell cultures for this purpose is more sensitive than the inoculation of monkeys, and therefore the appropriate change for the detection of residual virulent virus by cell cultures has been made. The Committee was informed that there is a pressing need to make this purified and concentrated vaccine more readily available.

The Committee noted the proposed revisions to the Requirements for Poliomyelitis Vaccine (Inactivated) (WHO/BS/79.1242 Rev. 2). The Committee adopted the revised Requirements for Poliomyelitis Vaccine (Inactivated) and agreed that they should be annexed to this report (Annex 2).

Annex 2

REQUIREMENTS FOR POLIOMYELITIS VACCINE (INACTIVATED)

(Requirements for Biological Substances No.2)
(Revised 1981)

PART C. REQUIREMENTS FOR HUMAN DIPLOID CELLS USED FOR VIRUS VACCINE PRODUCTION

The following requirements are applicable to the cell substrate for virus vaccine production where production is based on a cell seed system. The tests so far have been formulated for human diploid cells, but any cell bank and cell seed system shall comply with similar requirements as appropriate.

1. DEFINITIONS

1.1 Terminology

Cell seed. A quantity of cells derived from a single human tissue stored frozen at -70°C or below in aliquots, one or more of which would be used for the production of a manufacturer's working cell bank.

Manufacturer's working cell bank (MWCB). A quantity of cells derived from a single ampoule of the cell seed and of uniform composition stored frozen at -70°C or below in aliquots, one or more of which would be used for the production of each single harvest.

> In normal practice such a seed culture (or ampoule) is issued to manufacturers at or near the eighth population doubling level (PDL). This is expanded by serial subculture up to a PDL selected by the manufacturer, at which point the cells are combined into one or more pools and preserved cryogenically to form the MWCB. One or more of such ampoules from a pool would be used for the production of a single harvest.

Production cell culture. A collection of cell cultures at the population doubling used for virus growth that have been derived from a single ampoule of the MWCB.

2. GENERAL MANUFACTURING REQUIREMENTS

The general requirements contained in the revised Requirements for Biological Substances No.1 (General Requirements for Manufacturing Establishments and Control Laboratories) (2, page 11) shall apply, with the addition of the following directive:

No cell cultures other than those approved by the national control authority for the production of appropriate vaccine shall be introduced or handled in the production area.

3. PRODUCTION CONTROL

3.1 Cell seed

The use of human diploid cell cultures for vaccine manufacture shall be based on the cell seed system. Early population doubling of diploid cell cultures shall be subcultured to a population doubling that is convenient for the preparation of a cell seed.

The cell seed used for the production of virus vaccine shall be that approved by and registered with the national control authority. The accepted cell strain from which the cell seed has been derived shall have been characterized with respect to genealogy, growth characteristics, genetic markers (HLA), virus susceptibility, storage conditions, and karyology, and it shall have been shown, by tests in animals, eggs, and cell culture to be free from detectable adventitious agents.

These data shall be made available to the national control authority.

Each manufacturer shall show, to the satisfaction of the national control authority, that the cell substrate propagated from the accepted cell strain and laid down as a working cell bank conforms with the tests outlined in this section for freedom from extraneous agents, by tests in animals and eggs, for lack of tumorigenicity, for normal karyology throughout approximately the first two-thirds of its normal life-span, and for identity.

> In some countries the cells are examined also by ultra-thin, sections and negative staining under the electron microscope.

3.1.1 *Tests in animals and eggs for extraneous agents*

The tests in animals for extraneous agents shall include the inoculation of each of the following groups of animals with cells by the intramuscular route, using a least 10^7 viable cells divided equally between the animals in each group:

- 2 litters of suckling mice, comprising at least 10 animals, less than 24 h old;
- 10 adult mice;
- 5 guinea-pigs; and
- 5 rabbits.

At least 10^6 viable cells shall be injected also into the allantoic cavity of each of 10 embryonated chicken eggs 9-11 days old.

The animals shall be observed for at least 4 weeks and the embryonated chicken eggs shall be examined after not less than 3 days. Any animals that are sick or show any abnormality shall be investigated to establish the cause of illness. The allantoic fluids shall be tested with guinea-pig and chick or other avian red cells for the presence of haemagglutinins.

The cells are suitable for vaccine production if at least 80% of the animals or eggs inoculated with the cells remain healthy and survive the observation period and none of the animals or eggs shows evidence of the presence in the cell cultures of any extraneous agent.

3.1.2 *Tumorigenicity*

The cells at the production level shall also be shown to be free from tumorigenicity by appropriate tests in animals approved by the national control authority.

Particularly for new non-continuous cell strains, some control authorities may wish to consider applying the tests for tumorigenicity as outlined in Part D, section 3.1.2. For the diploid cell strains that have been in use for many years, however, the tests suggested below have been shown to be satisfactory.

Suitable tests using immunosuppressed animals may be made as follows. Approximately 10^6 viable cells obtained from cultures at the same passage level as those used for vaccine production are injected into (a) newborn mice or hamsters treated with antilymphocyte serum; or (b) athymic mice (nude nu/nu genotype); or (c) thymectomized mice irradiated and bone-marrow reconstituted (T -B +). Some of the same group of animals should be inoculated with a similar dose of HeLa or KB cells and it should be shown that tumour formation is caused by the inoculation of the neoplastic tissue, thus demonstrating the ability of the strain of animals to give rise to tumours. The animals should be observed for not less than 3 weeks. Any other test using animals treated with immunosuppressive agents and with equal sensitivity to neoplastic cells may be used.

Only those cell seeds shown not to be tumorigenic shall be used.

3.1.3 *Chromosomal characterization and monitoring*

3.1.3.1 *Chromosomal characterization.* At least 4 samples shall be examined as described in Part C, section 3.1.3.2, at approximately equal intervals over the life-span of the cell line during serial cultivation. Each sample shall consist of 1000 metaphase cells.

It is also recommended that photographic reconstruction should be employed in the preparation of chromosome-banded karyotypes of 50 metaphase cells per 1000 cell sample using either G-banding or Q-banding techniques. This constitutes 5 % of the sample and no specific limits for acceptability are yet recommended. The incidence of karyotypic abnormalities (pseudo-diploidy, inversions, translocations, etc.) that are detectable with the greater resolution provided by banding should be evaluated when a larger database than is at present available has been accumulated.

3.1.3.2 *Chromosomal monitoring - preparation and testing.* For the determination of the general character of each pool in the MWCB, a minimum of 500 cells in metaphase shall be examined at the production level or at any passage thereafter for frequency of polyploidy and for exact counts of chromosomes, frequency of breaks, structural abnormalities, and other abnormalities, such as despiralization or marked attenuations of the primary or secondary constrictions.

For vaccine production, examination of the cells is usually made between the 27th and 33rd population doubling. The national control authority should determine the level of cell population doubling allowable.

For cells examined in metaphase the upper limits[4] of acceptability (upper fiducial limits at 95 % (Poisson)) for abnormalities are: for 1000- and 500-cell samples, as follows:

Abnormality	1000 cells	500 cells
Chromatid and chromosome breaks	47	26
Structural abnormalities	18	10
Hyperploidy	8	5
Hypoploidy	180	90
Polyploidy	30	17

[4]These upper limits are based on extensive experience with the examination of WI-38 and MRC5 cells reported to and examined by the ad hoc Committee on Karyological Controls of Human Cell Substrates, which met in 1978 at Lake Placid, NY, USA. These values will not necessarily be applicable if another human cell strain is used.

All cells showing abnormalities shall be subjected to detailed examination, and records shall be maintained of the detailed criteria applied to particular abnormalities evaluated in the karyotype analysis.

Permanent stained slide preparations of the chromosome monitoring of the working cell bank pool, or photographs of these, shall be maintained as part of the record of the batch of vaccine for monitoring successive batches made from that cell pool.

It is desirable that a portion of the sample of pooled cell substrate removed from the culture vessel should be stored frozen so as to retain viability. This would be available for future reference for karyology or for any other purpose relating to the batch of vaccine.

Only those cell pools of the MWCB that have normal karyology shall be used for vaccine production.

3.1.4 *Identity test of the cells*

In some countries tests for characterizing HLA surface antigens are carried out in addition to chromosome monitoring.

3.2 Production of cell culture

A cell sample, equivalent to at least 500 ml of the cell suspension of the concentration employed for seeding the vaccine production cultures shall be used to prepare control cell cultures.

In some countries in which the technology of large-scale production has been developed the national control authority should determine the size of the sample of cells to be examined, the time at which the control cells should be taken from the production cultures, and the monitoring of the control vessels.

The treatment of cells set aside as control material shall be similar to that of the production cell culture but they shall remain uninoculated as control cultures for the detection of extraneous viruses.

These control cell cultures shall be incubated under similar conditions as the inoculated cultures for at least two weeks or until the time of the last harvest of the production cultures, whichever is the later, and shall be examined during this period for evidence of cytopathic changes. For the test to be valid, not more than 20 % of the control cell cultures may be discarded for nonspecific, accidental reasons.

At the end of the observation period, the control cell cultures shall be examined for degeneration caused by an infectious agent. If this examination or any of the tests required in this section show evidence of the presence in a control culture of any adventitious agent, the virus grown in the corresponding inoculated cultures shall not be used for vaccine production.

3.2.1 *Test for haemadsorbing viruses*
At the end of the observation period cells comprising 25% of the control cells shall be tested for the presence of haemadsorbing viruses using guinea-pig red cells. If the guinea-pig red cells have been stored, the duration of storage shall not have exceeded 7 days and the temperature of storage shall have been in the range of 2-8° C.

> In some countries the national control authority requires that tests for haemadsorbing viruses be made in addition using other types of red cells including those from humans (blood group 0), monkeys, and chickens (or other avian species). The cultures should be examined at 3-5 days and again at 12 days. All tests should be read after incubation for 30 min at 0-4°C and again after a further incubation for 30 min at 20-25° C. The test with monkey red cells should be read once more after yet another incubation for 30 min at 34-37° C.

3.2.2 *Tests for other extraneous agents*
At the time of each harvest of the production cultures and 14 days after the day of inoculation of the production cultures with seed lot virus, a sample of the pooled fluids shall be taken at each period of collection from each group of control cultures. 10 ml of each pool shall be tested in the same cells, but not the same batch of cells, as that used for the production of virus growth, and additional 10 ml samples of each pool shall be tested in human cells and at least one other sensitive cell system.

The inoculated cultures shall be incubated at a temperature of 35-37° C and shall be observed for a period of at least 14 days.

> Some national control authorities require that at the end of this observation period a subculture should be made in the same cell culture system and observed for at least 7 days. Furthermore, some national control authorities require that these cells should be tested for the presence of haemadsorbing viruses.

For the tests to be valid, at least 80% of the culture vessels should be available and suitable for evaluation at the end of the respective test periods.

If any cytopathogenic changes occur due to extraneous agents in any of the cultures the virus harvests produced from the batch of cells from which the control cells were taken shall be discarded.

3.2.3 *Identity test*
At the production level the cells shall be identified as human by tests approved by the national control authority.

> Suitable tests are isozymes analysis, HLA, or other immunological tests or karyotyping of at least one metaphase spread of chromosomes.

PART D.

REQUIREMENTS FOR CONTINUOUS CELL LINES USED FOR INACTIVATED VIRUS VACCINE PRODUCTION

The following requirements are applicable to the cell substrate for virus vaccine production where production is based on a cell seed system from a continuous cell line.

1. DEFINITIONS

1.1 Terminology

Cell seed. A quantity of cells derived from a normal tissue and stored frozen at -70°C or below in aliquots, one or more of which would be used for the production of a manufacturer's working cell bank.

Manufacturer's working cell bank (MWCB). A quantity of cells derived from one or more ampoules of the cell seed and of uniform composition stored frozen at -70°C or below in aliquots, one or more of which would be used for the production of each single harvest.

> In normal practice such a seed culture (or ampoules) is expanded by serial subculture up to a passage number selected by the manufacturer, at which point the cells are combined into one pool and preserved cryogenically to form the MWCB. One or more of the ampoules from such a pool would be used for the production of a single harvest.

Production cell culture. A collection of cell cultures at the passage number used for virus growth that have been derived from one or more ampoules of the MWCB.

2. GENERAL MANUFACTURING REQUIREMENTS

The general requirements contained in the revised Requirements for Biological Substances No.1 (General Requirements for Manufacturing Establishments and Control Laboratories) shall apply with the addition of the following directive:

No cell cultures other than those approved by the national control authority for the production of appropriate vaccine shall be introduced or handled in the production area.

3. PRODUCTION CONTROL

3.1 Cell seed and/or manufacturer's working cell bank

The utilization of continuous cell lines for vaccine manufacture shall be based on the cell seed system. A passage of a continuous cell line shall be subcultured to a passage number which is convenient for the preparation of a cell seed.

The cell seed used for the production of virus vaccine shall be that approved by and registered with the national control authority. The accepted continuous cell line from which the cell seed and/or the MWCB have been derived shall have been characterized with respect to genealogy, growth characteristics, immunological markers, virus susceptibility, and storage conditions, and it shall have been shown, by tests in animals, eggs and cell culture, to be free from detectable adventitious agents. In some countries karyology is also required.

These data shall be made available to the national control authority.

Each manufacturer shall show, to the satisfaction of the national control authority, that the cells intended as the virus substrate, propagated from the accepted continuous cell line and laid down as the MWCB conform with the test outlined in this section for freedom from extraneous agents by tests in animals and eggs (see Part D, section 3.1.1) and for lack of tumorigenicity. In addition, the test for tumorigenicity shall be repeated on the cells at the passage level used for vaccine production.

> In some countries the cells are examined by electron microscopy to establish their ultrastructural characteristics. The cells may also be tested for the presence of retroviruses after activation with agents such as bromodeoxyuridine (BUDR), by examining electron micrographs for virus particles and/or by performing assays for viral reverse transcriptase.

3.1.1 *Tests in animals and eggs for extraneous agents*

The tests in animals for extraneous agents should include the inoculation of each of the following groups of animals with cells by the intramuscular route, using at least 10^7 viable cells-divided equally between the animals in each group:

2 litters of suckling mice, comprising at least 10 animals, less than 24 h old;
10 adult mice;
5 guinea-pigs; and
5 rabbits.

At least 10^6 viable cells shall be injected also into the allantoic cavity of each of 10 embryonated chicken eggs 9-11 days old.

The animals shall be observed for at least 4 weeks and the embryonated chicken eggs shall be examined after not less than 3 days incubation. Any animals that are sick or show any abnormality shall be investigated to establish the cause of illness. The allantoic fluids shall be tested with guinea-pig and chick or other avian red cells for the presence of haemagglutinins.

The cells are suitable for vaccine production if at least 80% of the animals or eggs inoculated with the cells remain healthy and survive the observation period and none of the animals or eggs shows evidence of the presence in the cell cultures of any extraneous agent.

3.1.2 *Tests for tumorigenicity.*

Cells from the MWCB at the passage used for the virus production, or up to 10 passages thereafter, shall be shown to be non-tumorigenic in a test approved by the national control authority. Such a test must show a clear difference between the continuous cell line and a reference preparation of HeLa cells (information concerning the sources of suitable HeLa cells may be obtained from Chief, Biologicals, WHO, 1211 Geneva 27, Switzerland). In an appropriate *in vivo* test, 10^6 viable HeLa cells when given by the subcutaneous or intramuscular route shall produce progressively growing tumours in at least 9 out of 10 animals of which one or more must show evidence of metastases, while the 10^6 cells of an acceptable continuous cell line shall produce neither progressively growing tumours nor metastases.

The systems shown to be suitable for this test include:
(a) newborn mice, rats or hamsters that have been treated with antithymocyte serum or globulin,
(b) thymectomized and irradiated mice that have been reconstituted with bone marrow from healthy mice,
(c) chick embryo skin organ cultures.

A suitable test using newborn animals treated with antithymocyte serum is to inoculate at least 20 animals with 0.1 ml of potent serum within 24 hours of birth. The injection is given either by the intramuscular or subcutaneous route and is repeated on days 2, 7, and 14 of life. A potent antithymocyte serum or globulin is one that suppresses the immune mechanisms of the growing animals to the extent that the subsequent inoculum of 10^6 HeLa cells regularly produces tumours and metastases.

Also on the day of birth the two groups of 10 newborn animals that have been given the antithymocyte serum are given either 10^6 viable HeLa cells or 10^6 viable cells of the continuous cell line by the subcutaneous route at any site at which developing tumours can be palpated (the base of the neck or the abdomen are suitable sites). The animals are observed for 21 days for evidence of nodule formation at the site of injection and measurements are made at suitable times to determine whether there has been progressive growth.

At the end of the observation period all animals from both groups are sacrificed and examined for gross evidence of tumour formation at the site of injection and in other organs, such as the lymph nodes, lungs, kidneys, and liver. All tumour-like lesions are examined histopathologically. In addition, since some cell lines may form metastases without evidence of local tumour growth, the lungs and regional lymph nodes of all animals should be examined histopathologically.

For the purposes of this requirement, a progressively growing tumour is defined as a palpable nodule that increases in diameter over the 21-day observation period and that shows viable and mitotically active inoculated cells when examined histopathologically. The presence of microscopically viable cells without gross nodule formation should not be considered a progressively growing tumour; in addition, the presence of microscopically viable cells in association with a stationary or regressing nodule should not be considered a progressively growing tumour.

In addition, some countries test the cells for tumorigenicity in animals of the homologous species from which the cells were derived. Such tests would include immunosuppression of the animals with species-specific antithymocyte serum, inoculation of candidate cells and control tumour cells, observation for at least 3 weeks, and the histopathological examination of the inoculation sites as well as any metastatic lesions.

In some countries an *in vitro* test is permitted to demonstrate the freedom from tumorigenicity provided that the test has been shown to be as sensitive as a test in animals.

A suitable test using *organ cultures of chick embryonic skin* is to inoculate 10^6 HeLa cells or 10^6 cells from an acceptable continuous cell line on to organ cultures of chick embryonic skin for 3 days. At the end of this period, each culture is processed for histological evaluation and scored for cell growth and invasion. The reference HeLa cells should show extensive mitotic activity as well as extensive invasion into the chick substrate, while the continuous cell lines will show little or no invasion. In addition, secondary cell cultures derived from the same tissue as the continuous cell line may be tested in this system to provide guidance in interpreting invasive and mitotic activity.

3.1.3 *Identity test of the cells*

The MWCB shall be identified by a method approved by the national control authority.

The tests that may be used are karyology, isoenzymes analysis, and/or immunological markers.

3.2 Production of cell culture

A cell sample equivalent to at least 500 ml of the cell suspension at the concentration employed for seeding the vaccine production cultures shall be used to prepare control cultures.

In some countries in which the technology of large-scale production has been developed, the national control authority should determine the size of the sample of cells to be examined, the time at which the control cells should be taken from the production cultures, and the monitoring of the control vessels.

The treatment of cells set aside as control material shall be similar to that of the production cell cultures but they shall remain uninoculated as control cultures for the detection of extraneous viruses.

These control cell cultures shall be incubated under similar conditions to the inoculated cultures for at least 2 weeks or until the time of the last harvest of the production cultures, whichever is the later, and shall be examined during this period for evidence of cytopathic changes. For the test to be valid, not more than 20% of the control cell cultures may be discarded for nonspecific, accidental reasons.

At the end of the observation period, the control cell cultures shall be examined for degeneration caused by an infectious agent. If this examination or any of the tests required in this section shows evidence of the presence in a control culture of any adventitious agent, the virus grown in the corresponding inoculated cultures shall not be used for vaccine production.

3.2.1 *Test for haemadsorbing viruses*

At the end of the observation period cells comprising 25% of the control cells shall be tested for the presence of haemadsorbing viruses using guinea-pig red cells. If the guinea-pig red cells have been stored, the duration of storage shall not have exceeded 7 days and the temperature of storage shall have been in the range of 2-8° C.

3.2.2 *Tests for other extraneous agents*

At the end of the observation period a sample of the pooled fluids from each group of control cultures shall be tested for extraneous agents. 10 ml of each pool shall be tested in the same cells, but not the same batch of cells, as that used for the production of virus growth and additional 10 ml samples of each pool shall be tested in human cells and at least one other sensitive cell system.

The inoculated cultures shall be incubated at a temperature of 35-37° C and shall be observed for a period of at least 14 days.

For the tests to be valid, at least 80% of the culture vessels shall be available and suitable for evaluation at the end of the respective test periods.

If any cytopathogenic changes occur due to extraneous agents in any of the cultures, the virus harvests produced from the batch of cells from which the control cells were taken shall be discarded.

3.2.3 *Identity test*

At the production level the cells shall be identified as the species of origin by tests approved by the national control authority.

Suitable tests are isozymes analysis or other immunological tests, or karyology.

3.2.4 *Purity of virus prepared on a continuous cell line*

The virus grown in a continuous cell line shall be purified by a process that has been approved by the national control authority and has been shown to give consistent results.

Tests used to demonstrate the degree of purity achieved shall also be approved by the national control authority.

For poliomyelitis vaccine (inactivated) the purification process shall be shown to reduce consistently the level of cellular DNA from that of the initial virus harvest by a factor of at least 10^8. (see Part A, section 3.4.3.1).

AUTHORS

The first draft of these revised Requirements for Poliomyelitis Vaccine (Inactivated) was prepared by the following WHO consultants and staff members:

Dr F.A. Assaad, Virus Diseases, WHO, Geneva, Switzerland
Dr J. Furesz, Director, Bureau of Biologics, Drugs Directorate, Ottawa, Canada (*Consultant*)
Professor W. Hennessen, Berne, Switzerland (*Consultant*)
Dr G. Laurence, Connaught Laboratories Ltd., Willowdale, Ontario, Canada (*Consultant*)
Dr D.I. Magrath, Division of Viral Products. National Institute for Biological Standards and Control, London, England (*Consultant*)
Dr R. Mauler, Behringwerke AG, Marburg, Federal Republic of Germany (*Consultant*)
Dr J. Peetermans, Biological Division, Smith Kline - RIT, Rixensart, Belgium (*Consultant*)
Dr F.T. Perkins, Chief, Biologicals, WHO, Geneva, Switzerland

Dr J.C. Petricciani, Clinical Research, Bureau of Biologics, Food and Drug Administration, US Public Health Service, Bethesda, MD, USA (*Consultant*)
Dr J.D. van Rarnshorst, Biologicals, WHO, Geneva, Switzerland
Dr A.L. van Wezel, National Institute for Public Health; Bilthoven, Netherlands (*Consultant*)

The second revision that took into account the amendments necessary for the accommodation of the use of continuous cell lines was prepared by the following WHO consultants and staff members:

Dr A.J. Beale, The Wellcome Research Laboratories, Beckenham, England (*Consultant*)
Professor S.G. Dzagurov, Tarasevic State Institute, Moscow, USSR (*Consultant*)
Dr J. Furesz, Bureau of Biologics, Drug Directorate, Ottawa, Canada (*Consultant*)
Dr H. Lundbeck, The National Bacteriological Laboratory, Stockholm, Sweden (*Consultant*)
Dr R. Netter, Director-General, National Health Laboratory, Paris, France (*Consultant*)
Dr A. Nicolas, Mérieux Institute, Lyons, France (*Consultant*)
Dr E. Pearson, Connaught Laboratories Ltd., Ontario, Canada (*Consultant*)
Dr J. Peetermans, Biological Division, Smith Kline-RIT, Rixensart, Belgium (*Consultant*)
Dr F.T. Perkins, Chief, Biologicals, WHO, Geneva, Switzerland
Dr J. Pettriciani, Clinical Research, Bureau of Biologics, Food and Drug Administration, US Public Health Service, Bethesda, MD, USA (*Consultant*)
Dr L. Peyron, Mérieux Institute, Lyons, France (*Consultant*)
Dr J.D. van Ramshorst, Biologicals, WHO, Geneva, Switzerland
Dr Jonas E. Salk, The Salk Institute for Biological Studies, San Diego, CA, USA (*Consultant*)
Dr G. van Steenis, National Institute for Public Health, Bilthoven, Netherlands (*Consultant*)
Dr A. van Wezel, National Institute for Public Health, Bilthoven, Netherlands (*Consultant*)
Dr von Seefried, Connaught Laboratories Ltd., Willowdale, Ontario, Canada (*Consultant*)

ACKNOWLEDGEMENTS

Acknowledgements are due to the following experts for their comments and advice and for supplying additional data relevant to these requirements:
Dr S.C. Arya, New Delhi, India
Dr H.D. Brede, Paul Ehrlich Institute, Frankfurt-am-Main, Federal Republic of Germany
Dr V. Davey, Technical Director, Commonwealth Serum Laboratories, Parkville, Victoria, Australia
Mr I. Davidson, Head, Biological Products and Standards Department, Central Veterinary Laboratory, Weybridge, Surrey, England

Professor E. Geissler, Academy of Sciences of the German Democratic Republic, Research Centre for Molecular Biology and Medicine, Central Institute for Molecular Biology, Berlin
Dr A. Gray, Merck, Sharp & Dohme, West Point, Pennsylvania, USA
Dr H. Gruschkau, Behringwerke, Marburg, Federal Republic of Germany
Dr C. Guthrie, Production Director, Commonwealth Serum Laboratories, Parkville, Victoria, Australia
Dr L. Hayflick, Children's Hospital Medical Center, Oakland, CA, USA
Dr H.P. Lansberg, National Institute for Public Health, Bilthoven, Netherlands
Dr H. Mirchamsy, Associate Director, Razi State Institute of Serum and Vaccine Production, Teheran, Iran
Professor F. Pocchiari, Director-General, Higher Institute of Health, Rome, Italy
Dr E.J. Ruitenberg, National Institute for Public Health, Bilthoven, Netherlands
Dr J.J. Salaun, Pasteur Institute, Dakar, Senegal
Dr J.R. Thayer, National Biological Standards Laboratory, Canberra City, Australia
Dr W. Aeg. Timmerman, De Bilt, Netherlands
Dr A.L. van Wezel, National Institute for Public Health, Bilthoven, Netherlands
Mr J. Withell, Chief, Viral Products Branch, National Biological Standards, Parkville, Victoria, Australia

REFERENCES

1. WHO Technical Report Series, No. 178, 1959.
2. WHO Technical Report Series, No. 323, 1966.
3. WHO Technical Report Series, No. 530, 1973.

This report contains the collective views of an international group of experts and does not necessarily represent the decisions or the stated policy of the World Health Organization

Acceptability of cell substrates for production of biologicals

Report of a WHO Study Group

World Health Organization
Technical Report Series (TRS)
747

World Health Organization, Geneva **1987**

The World Health Organization is a specialized agency of the United Nations with primary responsibility for international health matters and public health. Through this organization, which was created in 1948, the health professions of some 165 countries exchange their knowledge and experience with the aim of making possible the attainment by all citizens of the world by the year 2000 of a level of health that will permit them to lead a socially and economically productive life.

By means of direct technical cooperation with its Member States, and by stimulating such cooperation among them, WHO promotes the development of comprehensive health services, the prevention and control of diseases, the improvement of environmental conditions, the development of health manpower, the coordination and development of biomedical and health services research, and the planning and implementation of health programmes.

These broad fields of endeavour encompass a wide variety of activities, such as developing systems of primary health care that reach the whole population of Member countries; promoting the health of mothers and children; combating malnutrition; controlling malaria and other communicable diseases including tuberculosis and leprosy; having achieved the eradication of smallpox, promoting mass immunization against a number of other preventable diseases; improving mental health; providing safe water supplies; and training health personnel of all categories.

Progress towards better health throughout the world also demands international cooperation in such matters as establishing international standards for biological substances, pesticides, and pharmaceuticals; formulating environmental health criteria; recommending international nonproprietary names for drugs; administering the International Health Regulations; revising the International Classification of Diseases, Injuries, and Causes of Death; and collecting and disseminating health statistical information.

Further information on many aspects of WHO's work is presented in the Organization's publications.

*   *
  *

The *WHO Technical Report Series* makes available the findings of various international groups of experts that provide WHO with the latest scientific and technical advice on a broad range of medical and public health subjects. Members of such expert groups serve without remuneration in their personal capacities rather than as representatives of governments or other bodies. An annual subscription to this series, comprising 12 to 15 such reports, costs Sw. fr. 85.-.

ISBN 92 4 120747 7

ISSN 0512-3054

PRINTED IN SWITZERLAND

7122 - Schüler S.A. - 7000

CONTENTS

WHO STUDY GROUP ON BIOLOGICALS
Geneva, 18-19 November 1986

Members

Professor G. L. Ada, Head of Microbiology Department, John Curtin School of Medical Research, Australian National University, Canberra, Australia (*Rapporteur*)
Dr J. Chermann, Pasteur Institute, Paris, France
Dr M.G. Deo, Director, Indian Cancer Research Centre, Bombay, India
Professor M.A. Epstein, John Radcliffe Hospital, Oxford, England
Professor H. Harris, Head, Sir William Dunn School of Pathology, Oxford, England
Dr H. Koprowski, Director, Wistar Institute, Philadelphia, PA, USA
Dr V.A. Laskevic, Deputy Director, Institute of Poliomyelitis and Viral Encephalitides, Moscow, USSR
Dr D.R. Lowy, Chief, Laboratory of Cellular Oncology, National Cancer Institute, Bethesda, MD, USA
Dr M.A. Martin, Chief, Laboratory of Molecular Microbiology, National Institute of Allergy and Infectious Diseases, Bethesda, MD, USA
Professor T. Matuhasi, Okinaka Memorial Institute for Medical Research, Tokyo, Japan
Dr C. Morel, Director, Oswaldo Cruz Institute, Rio de Janeiro, Brazil
Dr F. Robbins, Case Western Reserve University, Cleveland, OH, USA (*Chairman*)
Dr R. Sager, Dana-Farber Cancer Institute, Boston, MA, USA
Dr C. R. Steinman, Department of Medicine, State University of New York, New York, NY, USA
Professor K. Takatsuki, University of Kumamoto, Kumamoto, Japan
Professor A.J. van der Eb, Sylvius Laboratory, Leiden, The Netherlands
Dr Wu Min, Cancer Institute, Chinese Academy of Medical Sciences, Beijing, China
Dr D. Zewdie, National Research Institute of Health, Addis Ababa, Ethiopia (*Vice-Chairman*)
Dr H. zur Hausen, German Cancer Research Centre, Heidelberg, Federal Republic of Germany

Secretariat

Dr J. Doehmer, Trinity College, Dublin, Ireland (*Temporary Adviser*)
Dr A.I. Kingsman, University of Oxford, Oxford, England (*Temporary Adviser*)
Dr M.P. Moyer, University of Texas Health Science Center, San Antonio, TX, USA (*Temporary Adviser*)
Dr J.C. Petricciani, Chief, Biologicals, WHO, Geneva, Switzerland (*Secretary*)
Dr A.J. Strain, Department of Paediatrics, University of Sheffield, England (*Temporary Adviser*)

ACCEPTABILITY OF CELL SUBSTRATES FOR PRODUCTION OF BIOLOGICALS

Report of a WHO Study Group on Biologicals

A WHO Study Group on Biologicals met in Geneva on 18 and 19 November 1986. The meeting was opened on behalf of the Director-General by Dr S. Litvinov, Assistant Director-General.

1. BACKGROUND AND INTRODUCTION

The acceptability of cell substrates for vaccine production has been a subject of controversy ever since the 1950s, when the decision was made that primary cell cultures of non-human primate origin were acceptable for the production of certain vaccines such as poliomyelitis vaccine. This decision set an important precedent, which has had an impact on the consideration of alternative cell systems. For example, it took a decade or more for human diploid cell substrates to be accepted by certain national control authorities, even though all of the available scientific evidence indicated that products derived from them were safe to use.

The first reconsideration of the acceptability of substrates other than primary and diploid cells occurred in 1978 at Lake Placid, USA, when human lymphoid (lymphoma-derived) cells were proposed as a source of α-interferon. The major reason for considering a human malignant cell as a substrate for the production of interferon was that the product could be generated in very large quantities and in such a way that the manufacturing process could ensure the elimination of cellular contaminants that might be of concern. This research and development effort laid the groundwork for methods that might be used to ensure the safety of products derived from a variety of other continuous cell lines.

As advances were made during the past 10 years in basic biological research, and as it became clear that recombinant DNA techniques would provide an opportunity to develop biological products that it had not previously been possible to manufacture, the usefulness of continuous cell lines as substrates became apparent. The development of hybridoma technology also heightened interest in the use of such cell lines. As a result, a series of meetings was held in Europe and North America to discuss the subject further. Products for clinical trials were then developed in cell lines by various groups, and certain national control authorities began to approve the experimental use of these products. More recently, national control authorities have also licensed certain products (e.g., lymphoid interferon, monoclonal antibodies, and inactivated rabies and poliomyelitis vaccines). In addition, the WHO Expert Committee on Biological Standardization approved the use of non-tumorigenic and virus-free continuous cell lines for the production of inactivated poliomyelitis vaccine in 1981. In fact, the general trend since 1978 has been towards the acceptance of continuous cell lines for the production of various biologicals. However, when a group of consultants met to consider the development of WHO Requirements for hepatitis B vaccine produced by recombinant DNA techniques in continuous cell lines, they suggested that it would be useful for WHO to convene a group of experts to consider issues associated with the use of eukaryotic and prokaryotic cells in the production of biologicals, with particular emphasis on the potential risks from contaminating DNA, in order to provide international guidance on the acceptability of various cell substrates and production strategies. The Director-General therefore convened the Study Group on Biologicals to provide advice to WHO on these issues.

Background documents providing both experimental data and theoretical information were discussed and led to the formulation of the conclusions provided later in this report (section 3). A major concern was the long-term risk of malignancy represented by heterogeneous contaminating DNA, especially if it were to contain potentially oncogenic coding or regulatory sequences. This concern was especially real because many healthy people, including infants, might be vaccinated with, or otherwise receive, products derived from continuous cell lines.

There is some evidence for the safety of certain continuous cell lines as substrates for the production of biologicals. For example, an inactivated virus vaccine for foot and mouth disease has been prepared in the BHK-21 cell line, and it is estimated that more than 100 million doses have been administered to cattle over a 20-year period. Carcass inspection has failed to reveal any ill-effects attributable to the vaccine, suggesting that in the short term (2-4 years) this vaccine has been safe. Although clinical experience with biologicals produced for human use in continuous cell lines has been more recent, and therefore more limited, than experience with veterinary products, it was noted that about 19 million doses of inactivated poliomyelitis vaccine produced in Vero cells had already been administered to children in the three years since 1983.

2. ISSUES CONSIDERED BY THE STUDY GROUP

The issues were of two main types: firstly, the acceptability of developing a biological product in a new cell system when the same generic product is already being manufactured by an approved method (see *a* below); and secondly, the degree of risk associated with certain classes of possible contaminants in the product, including heterogeneous contaminating DNA (see *b-f* below); viruses (see *g* below); and transforming proteins (see *h* below).

The specific questions discussed by the Study Group were as follows:

- *(a)* What position should be taken on the acceptability of a product generated in a continuous cell line when an alternative production method exists in another cell system? For example, alpha-interferon can be produced in *Escherichia coli* as well as in human lymphoid cells; and hepatitis B vaccine can be produced from human plasma and in yeast, as well as in continuous cell lines.
- *(b)* Is there an absolute amount of DNA per dose of product (or accumulated dose for repetitively administered biologicals) below which the probability of a biological effect is so small as to be effectively zero? Would nuclease treatment enhance confidence in the safety of the product, or simply add another concern, perhaps even more significant than that related to the DNA?
- *(c)* What type of DNA should be the focus of attention in tests to quantify the amount of DNA in a product or in a given step in the manufacturing process? If regulatory

sequences are used in constructing the recombinant cell, should specific efforts be made to look for such sequences? Are highly repetitive and widely dispersed sequences, such as *Alu* in human DNA, adequate as a generic probe?

(d) Are viral sequences, if they exist in any given cell line, of special concern, or can they be dealt with in a generic sense with the rest of the cellular DNA?

(e) What are the experimental models of choice for assessing the transforming potential of DNA from cells being considered for use, and how long should the studies last?

(f) How valuable are validation studies of the ability of a manufacturing process to eliminate or inactivate unwanted viruses and DNA as compared with studies on the final product?

(g) Are the screening procedures now contained in WHO Requirements for the characterization of continuous cell lines adequate to detect viruses of concern?

(h) Should continuous cell lines be characterized regarding their ability to produce transforming proteins? What are the assay systems of choice? If transforming proteins are found, how significant is the risk to the recipient of the product and what levels would be acceptable?

(i) What criteria of purity should be applied to the final product to provide a reasonable assurance of acceptability?

3. DISCUSSION AND CONCLUSIONS

Several features make continuous cell lines particularly suitable as substrates for the production of a wide range of biologicals for human use. These include:

- low cost (relative to other mammalian substrates such as primary cells, which must be intensively tested at each harvest);
- avoidance of the use of primates as the source of primary cells for production;
- ability to prepare, standardize, and store cell seeds; ease of preparing suspension cultures on a very large scale;
- susceptibility to infection by a variety of viruses of human importance;
- relative ease of transfection with recombinant DNA plasmids and subsequent cloning of high-yield transformants;
- high likelihood of correct post-translation processing of mammalian proteins encoded by transfected DNA, which enhances the probability of correct conformational structure of the product;
- secretion (with or without genetic manipulation) of products into the medium.

These properties are important not only for the production of biologicals in developed countries, but could greatly facilitate the transfer of vaccine-producing capability to developing countries, which is an important goal of WHO.

There are, however, potential risks associated with the human use of biologicals produced in continuous cell lines. Before describing and analysing these risks, the Study Group reviewed general issues associated with the acceptance of new biologicals.

3.1 Risk assessment during research and development

There are several important steps in the progression of a novel approach from concept to practice. The first step is an evaluation of the risks that the new drug, vaccine, or cell substrate does or might pose to human beings. In most countries, if not all, preclinical safety data are required before the administration of a new drug or biological product to human subjects. Indeed, international as well as national ethical norms have been established to protect human subjects from unreasonable risks during the experimental phases of the development of new therapeutic and prophylactic products.

Before products derived from continuous cell lines were first approved for human use, different committees assessed the associated benefits and risks, taking into consideration not only the data in support of product safety but also the relative benefits of the new products in comparison to existing products. Before proceeding with human clinical studies, the committees evaluated the safety of each product on the basis of all the data available, taking into account the capacity of various steps in the manufacturing process to remove and inactivate potentially harmful contaminants. Some national control authorities also made a decision on the acceptability of proceeding with experimental studies in human subjects.

Those clinical study committees, and in some cases national control authorities as well, concluded that it was reasonable to proceed with human studies with a variety of products derived from continuous cell lines. They did not expect that the use of a continuous cell line would result in any real risk for the human subjects. It would be inimical to the basic concepts of medicine knowingly to impose the risk of cancer on patients with serious diseases or on healthy people, particularly since alternatives exist to many of the products that are now available from continuous cell lines or that are expected to be approved soon. The general view of the groups that have considered these issues so far is that, whatever the potential risks might have been, the data in support of the safety of the products justified their acceptance. Nevertheless, it is prudent to be alert to unexpected risks, and appropriate long-term follow-up studies of groups of recipients of these new products should be considered as they are introduced into general use. Taking this into consideration, the Study Group concluded that the basic decision on the safety of a product derived from a continuous cell line must be made at the point of approving the first clinical trial. That initial judgment needs to be modified only if new and relevant data subsequently come to light.

3.2 Products manufactured by alternative methods

The Study Group also considered the acceptability of using continuous cell lines for products such as α-interferon, poliomyelitis vaccine, and hepatitis B vaccine, which can be produced by alternative methods. Both the individual and society are continually faced with choices involving relative safety and benefits. The selection of acceptably safe options from among various alternatives has played a key role in the history of biological product development, and has included a consideration of issues such as which virus strains and which cell substrates to use for vaccine production.

The Study Group concluded that, when reliable data are provided in support of the safety of a product derived from a continuous cell line, that product should be considered acceptable. The existence of an approved manufacturing process for the same product in another cell system is not considered to be relevant to the acceptance or rejection of a product derived from a continuous cell line. Every product should stand on its own merits regarding safety and efficacy.

3.3 Potential risks associated with biologicals produced in continuous cell lines
The main potential risks associated with the use of biologicals produced in continuous cell lines fall into three categories: heterogeneous contaminating DNA, viruses, and transforming proteins. A summary of the risk assessment for each follows. More comprehensive statements on heterogeneous contaminating DNA and transforming proteins are provided as Annexes 1 and 2, respectively.

3.3.1 *Heterogeneous contaminating DNA*
The Study Group concluded, on the basis of the experimental data available, that the risk associated with heterogeneous contaminating DNA in a product derived from a continuous cell line is negligible when the amount of such DNA is 100 pg or less in a single dose given parenterally. The assessment of the safety of any product with respect to DNA should take into consideration: (*a*) the elimination of the biological activity of DNA by various steps in the manufacturing process; and (*b*) the reduction in the amount of DNA during the purification of the product in the manufacturing process. A given product may be considered safe on the basis of reliable data on either or both of these operations. Because of the extremely low probability of a biological effect of 100 pg of heterogeneous contaminating DNA per dose of product given parenterally, nuclease treatment of products during manufacture would probably add more concerns that it would remove.

The use of special DNA sequences, such as viral regulatory sequences, in the construction of recombinant cells is considered acceptable because there is no evidence that such sequences would impose any additional risk beyond that of heterogeneous contaminating cellular DNA in general. Nevertheless, the manufacturing process should be validated to demonstrate that such sequences are not concentrated in the crude harvest or at other steps. It is expected, however, that recombinant DNA products will be highly purified and the amount of DNA per dose will be less than 100 pg; in some cases it may even be undetectable.

When a product is likely to be contaminated by DNA with important biological activity, detection of that DNA should be attempted by appropriate techniques, some of which are referred to in Annex 1. Newer techniques for detecting DNA are under development.

Viral inactivating agents have been used for many years in the preparation of safe and effective vaccines for poliomyelitis, rabies, and hepatitis B. Some data suggest that these viral inactivating agents may also destroy the biological activity of ONA, thus providing an additional level of confidence in the safety of these products even when the amount of DNA in a parenteral dose of the product is above 100 pg. However, the Study

Group concluded that more specific data were needed on the effects of these inactivating agents under manufacturing conditions in order to draw firm conclusions on their DNA-inactivating potential.

3.3.2 Viruses

The Study Group reviewed the potential risk to human recipients of products manufactured in cells containing viral agents. These agents may include complete viruses with known patterns of replication such as simian virus 40, virus particles such as the type A retroviruses that can be visualized by electron microscopy, and persisting viral genomes or parts of genomes, for example those of the hepatitis B and Epstein-Barr viruses. Cells may be divided into three risk categories with respect to their potential for carrying viral agents pathogenic for human beings:

High risk	Blood and bone marrow cells derived from human or non-human primates; caprine and ovine cells; hybridomas when at least one fusion partner is of human or non-human primate origin.
Medium risk	Mammalian non-haematogenous cells such as fibroblasts and epithelial cells.
Low risk	Human diploid cell lines and cells derived from avian tissues.

In making these risk distinctions, the following points were noted:

(*a*) Human lymphocytes and macrophages may carry latent viruses, such as human retroviruses, which may become activated when the cells are exposed to growth-stimulating factors during *in vitro* cultivation. The existence of simian retroviruses infectious for human cells places haematogenous cells from non-human primates in the high risk group. If primary human leukocytes are used for the production of interferon or other biologicals, the manufacturing process should be demonstrated to inactivate the known human retroviruses. In addition, it will be important that products manufactured in primate non-haematogenous cells be tested for viruses that are pathogenic in human subjects and are known to be harboured by the species from which the cells were originally derived. Caprine and ovine cells have been shown frequently to be contaminated with lentiviruses and "slow viruses" associated with subacute spongiform encephalopathies.

(*b*) Primary monkey kidney cells have been used to produce millions of doses of poliomyelitis vaccines over the past 20 years, and although latent viruses such as simian virus 40 were discovered in such cells, control measures were introduced to eliminate any risk associated with the manufacture of vaccines in cells containing those endogenous viruses. Additional controls may be needed as new viral agents are identified.

(*c*) Continuous lines of non-haematogenous cells from human and non-human primates may contain viruses or have viral genes integrated into their DNA. In either case, virus expression may occur under in vitro culture conditions.

(*d*) Rodent and avian tissues and cells are well known to harbour viruses, but there is no evidence for transmission of disease to human beings by products from these

sources. For example, large quantities of yellow fever, measles, and live influenza vaccines have been produced for many years in eggs that contain avian leukosis viruses, but there is no evidence that these products have had any harmful effects in their long history of use for human immunization. On the other hand, lymphocytic choriomeningitis virus and haemorrhagic fever viruses harboured by rodents have caused disease in humans by direct infection.
(*e*) Human diploid fibroblasts have been used for vaccine production for over 10 years, and although initial concern was expressed about the possibility of the cells containing a latent human virus, no evidence for such an agent has been found, and vaccines produced from this class of cell have proved to be safe.

Taking into account this classification of cells according to their potential for transmitting viruses pathogenic for human beings, the Study Group agreed that different degrees of concern, and therefore testing, were appropriate for products manufactured from the various types of cell mentioned.

Nevertheless, it was emphasized that when either diploid cell lines or continuous cell lines are used for production, a cell seed lot system must be used and the cell seed must be characterized as specified in the appropriate WHO Requirements. Efforts to identify viruses, viroids, and similar structures should constitute an important part of the characterization of such cell banks. For example, the possible presence of contaminating viroid-like RNA, such as the Delta agent, should receive attention and appropriate methods should be used for its detection.

When cell lines of rodent or avian origin are examined for the presence of viruses, the major emphasis in risk assessment should be placed on the results of studies in which transmission to target cells or animals is attempted. Risk to human recipients should not be assessed *solely* on ultrastructural evidence for the presence of viral agents in the cells.

There may be as yet undiscovered microbial agents for which there is no current evidence or means of detection. However, during the long history of use of animals, tissues, and cells to manufacture products for human administration, the rare cases of viral contamination detected have all been due to activation of known latent viruses, either in the starting material or in the product, rather than to "new" agents that transmitted disease to human beings.

The Study Group stressed the importance of validating the ability of a manufacturing process to eliminate and to inactivate those viruses that may pose a risk to human beings when cells or cell lines carrying such viruses are proposed for use in the manufacture of biologicals for human administration. As with heterogeneous contaminating DNA, a wide margin of safety should be established in any inactivation or purification process. In addition, cells from human beings or animals with diseases of unknown origin and animal cells that may contain "slow viruses" should not be used to produce biologicals for human administration.

3.3.3 *Transforming proteins*
The apparent risk from oncogene-encoded proteins is limited to growth factors, since they are the only such proteins that can exert their biological effects via the cell surface. Growth factors may be secreted by cells used to produce biologicals, but the risks from growth factors are limited, since their growth-promoting effects are usually transient and reversible, they do not replicate, and many of them are rapidly inactivated *in vivo*.

Growth factors do not ordinarily appear to be oncogenic. Given the levels at which known growth factors are secreted by cultured cells, it would be necessary for them to be concentrated from the culture medium before they would be expected to be biologically active *in vivo*. In exceptional circumstances, however, growth factors can contribute to oncogenesis, but even in these cases, the tumours apparently remain dependent upon continued administration of the growth factor.

In summary, the Study Group did not consider that the presence of contaminating known growth factors in the concentrations at which they are ordinarily to be found constitutes a serious risk in the preparation of biological products from continuous cell lines.

3.4 General conclusions
Although it is possible to estimate an upper limit of contamination of a final product with heterogeneous DNA, and although all reported experiments have indicated that picogram amounts of such DNA are biologically inactive in a variety of tests, the *complete* absence of DNA or its associated risk cannot be claimed for products derived from continuous cell lines any more than for products derived from primary and diploid cell cultures. However, the probability of DNA from continuous cell lines or other cell systems inducing malignancy and other diseases is considered to be extremely low. In addition, there are well-documented and effective methods available for the preparation of safe products. For example, using appropriate purification and inactivation procedures, it has been possible to prepare a safe hepatitis B vaccine from the blood of infected patients, although this source of material carries very high risks. Many national control authorities have approved such hepatitis B vaccines because reliable data show that the manufacturing process has resulted in safe and effective products which meet WHO Requirements.

The importance of validating the efficiency with which various steps in a manufacturing process inactivate and/or eliminate unwanted material such as cellular DNA and viruses was emphasized. Validating the ability of a process to yield a product with certain specifications and establishing the consistency of that process are essential in providing the basis for an acceptable biological derived from continuous cell lines. Once a process has been validated and consistency of production has been established, limited tests appropriate for each product should suffice, as has been the usual practice with biologicals in the past. It was noted that this is the approach proposed in the WHO Requirements for hepatitis B vaccine produced by recombinant DNA techniques.

The Study Group concluded that, in general, continuous cell lines are acceptable as substrates for the production of biologicals, but that differences in the nature of the products and in the characteristics of the manufacturing processes must be taken into account in making a decision on the acceptability of a given product. There is, therefore, no reason to exclude continuous cell lines from consideration as substrates for biological products. In this regard, there was agreement with the actions taken to date by committees in approving the use of various products derived from continuous cell lines only when they were satisfied that the manufacturing process in question yielded a product with no detectable risk attributable to the cell substrate.

In addition to those already available, a variety of other biologicals will eventually be produced in continuous cell lines. These can currently be grouped broadly into three classes: vaccines, biologically active proteins, and monoclonal antibodies. The dose and frequency of administration of these products may vary widely among the three major groups as well as within a group. This emphasizes the importance of taking into consideration the specific details of a product such as the dose, route of administration, and frequency of administration when assessing the ability of a manufacturing process to provide a safe product.

The risks from heterogeneous contaminating DNA were considered negligible for preparations given orally. For such products the principal requirement is the elimination of contaminating viruses and toxic proteins. If the manufacturing principles advocated by the Study Group for parenteral products are followed, the risk from oral administration of these products is greatly reduced.

4. RECOMMENDATIONS

1. WHO should promote establishment of several banks of cell seeds for continuous cell lines to assist Member States and manufacturers, when requested, in developing working cell banks that conform to WHO Requirements for the characterization of continuous cell lines for the production of biologicals.
2. WHO should take all possible steps to encourage the replacement of biologicals derived from neural tissues with biologicals derived directly from cell culture, including continuous cell lines, or made by recombinant DNA techniques.
3. WHO should encourage and coordinate studies to determine the effect of various inactivating agents, such as propiolactone and formalin, on the biological activity of DNA.
4. National control authorities should consider establishing multidisciplinary groups to assist them in assessing the safety and acceptability of substances produced by modern biological techniques, since the issues are often new and complex, and require the collective wisdom of experts from several different disciplines.
5. The Study Group recognizes that the matters considered in this report, especially the malignant transformation of human cells, are fast-moving fields of research. The Study Group therefore recommends that the WHO Secretariat should keep a close watch on further developments, and requests the Director-General to convene groups of scientists as appropriate.

5. ACKNOWLEDGEMENTS

The Study Group acknowledges the valuable contribution to its work made by the following staff members of the World Health Organization and the International Agency for Research on Cancer: Dr F. Assaad, Director, Division of Communicable Diseases, WHO, Geneva, Switzerland; Dr J. Dunne, Chief, Pharmaceuticals, WHO, Geneva, Switzerland; Dr J. Esparza, Microbiology and Immunology Support Services, WHO, Geneva, Switzerland; Dr Y. Ghendon, Microbiology and Immunology Support Services, WHO, Geneva, Switzerland; Dr V. Grachev, Biologicals, WHO, Geneva, Switzerland; Dr R. Montesano, International Agency for Research on Cancer, Lyon, France; Dr Y. Pervikov, Microbiology and Immunology Support Services, WHO, Geneva, Switzerland; and Dr P. Sizaret, Biologicals, WHO, Geneva, Switzerland.

Annex 1

HETEROGENEOUS CONTAMINATING DNA

The major risk associated with heterogeneous contaminating DNA in biological preparations intended for human use is connected with its potential pathogenic activity. After consideration of the data available and of calculations of the tumour-inducing potential of residual DNA concentrations, the Study Group concluded that there is negligible risk from heterogeneous contaminating DNA at a concentration of 100 pg or less per parenteral dose. This value is based on experimental results obtained in tests in which the pathogenic activity of DNA sequences was measured in susceptible animals. The following DNA sequences were examined:

- DNA of oncogenic viruses, including polyoma virus, simian virus 40, adenovirus, and Rous sarcoma virus;
- cloned genomic DNA from hepatitis B virus;
- chromosomal DNA from tumour cells and cloned mutant e-ras genes.

DNA of oncogenic viruses, including polyoma virus, simian virus 40, adenovirus, and Rous sarcoma virus

Viral DNA injected into experimental animals is weakly oncogenic (Table 1). The results can be summarized as follows:

Polyoma virus.

Injection of 0.5-2 μg of polyoma virus DNA into newborn hamsters or rats resulted in the induction of tumours in 10-80% of the animals (*1-3*).

Simian virus 40

Subcutaneous injection of 1 μg or 2 μg of simian virus 40 DNA into newborn hamsters gave rise to sarcomas in 33-55% of the animals (*2*, *4*).

Adenoviruses

Doses of 3-5 μg of simian adenovirus 7 DNA were reported to induce tumours in newborn and 21-day-old hamsters (*5*, *6*). Injection into newborn hamsters of 4 μg of human adenovirus 12 DNA or an equivalent amount of the cloned transforming DNA fragment resulted in tumour formation in 2 out of 50 animals (*7*).

Rous sarcoma virus

Injection of 2 μg of a subgenomic proviral DNA fragment of Rous sarcoma virus containing the viral src gene into the wing web of chickens caused tumours in 65% of the animals (*8*). These tumours regressed after several weeks. These and other results indicate that DNA from oncogenic viruses is able to induce tumours in animals, but usually only when injected in doses between 1 and 10 μg.

Table 1. Tumorigenicity of tumour virus DNA in animals

DNA		Animals tested	Route of injection	Tumour induction[a]	Reference
Source	Amount (µg)				
Polyoma virus	0.5	Newborn hamster	i.p.	10% (5/52)	1
	1	Newborn hamster	s.c.	11% (2/22)	3
	2	Newborn rat	s.c.	60% (33/55)	2
	0.2	Newborn rat	s.c.	22% (2/9)	2
	2[b]	Newborn hamster	s.c.	80% (4/5)	2
Simian virus 40	1-2	Newborn hamster	s.c.	33% (11/33)	4
	2	Newborn hamster	s.c.	55% (4/7)	2
	2[c]	Newborn hamster	s.c.	53% (9/17)	2
	1-10 (sub- genomic DNA)	Newborn hamster	s.c.	0% (0/131)	11
Adenovirus					
Simian adenovirus 7	3	Newborn hamster	s.c.	28% (7/25)	5
	5	Newborn hamster	s.c.	30% (24/82)	6
	2.5	Newborn hamster	s.c.	7% (4/59)	6
Human adenovirus 12	4	Newborn hamster	s.c.	4% (2/50)	7
Rous sarcoma virus *(v-src)*	2	Chicken	Wing web	65% (11/17)	8

[a]Percentage of animals in which tumours developed: actual numbers are given in parentheses.

[b]This plasmid expressed the polyoma middle T protein only. The latency periods for tumour induction were 5-10 times longer than with wild-type polyoma DNA. The middle T gene did not induce tumours in newborn rats.

[c]A combination of the genes for simian virus 40 large T antigen and for the polyoma small T antigen.

Cloned genomic DNA from hepatitis B virus
Intrahepatic injection into chimpanzees of 5 ug of cloned DNA from hepatitis B virus caused hepatitis B, but intravenous injection of this DNA did not cause disease (*9*).

Chromosomal DNA from tumour cells and cloned mutant *c-ras* genes
Experiments with chromosomal DNA from tumour cells and with cloned activated *ras* genes have recently been conducted in several laboratories or are in progress (Table 2). So far, all tests have been negative, both with chromosomal DNA isolated from T24 bladder carcinoma cells and with the cloned activated H-*ras* or K-*ras* genes, at least during the observation periods used. The amounts of DNA injected varied from 10 to 500 μg for chromosomal DNA and from 2 to 50 μg for cloned activated ras genes (*3*, *10*).

Calculations concerning DNA concentration and tumour-inducing potential
Experiments have shown that 2 μg of DNA from polyoma virus or simian virus 40 can induce tumours in about 50% of animals tested. If 2 μg of DNA is defined as the "tumour-inducing dose", then a residual amount of 100 pg of tumour virus DNA in a biological preparation would correspond to

$$100/2x10^6 = 0.5 \times 10^{-4} \text{ tumour-inducing dose.}$$

The contaminating DNA, however, normally consists of chromosomal DNA and not of pure viral DNA. If the chromosomal DNA contains an activated oncogene and there is only one copy per genome, the oncogene will represent only $1/10^6$ of the total DNA.[1] Then 100 pg of heterogeneous contaminating chromosomal DNA would contain $100 \times 10^{-6} = 10^{-4}$ pg of activated oncogene, which corresponds to

$$10^{-4}/2x10^6 = 0.5x10^{-10} \text{ tumour-inducing dose.}$$

The risk associated with this amount of DNA is so small that it can be safely regarded as being negligible.

Several points should be considered in these risk estimations. Firstly, all calculations are based on the assumption that the risk factor for tumour induction decreases linearly with decreasing DNA concentration. This may not necessarily be correct, since an amount of DNA that has no measurable biological effect in a standard assay because it is present at too low a concentration may still have an effect under certain conditions or in certain organs or tissues. Secondly, it is not clear whether the risk associated with consecutive exposures to DNA will act in a cumulative way or not. Thirdly, the possibility should be considered that preparations of DNA that do not produce tumours in experimental systems may in human beings induce changes that could increase the incidence of tumours developing after long latent periods. Fourthly, experiments with short-lived animals do not permit assessment of the long-term effects of acquired DNA sequences.

[1] If the amount of genomic DNA per cell has a mass of 10 pg and an oncogene measures 10 kilobases (= 10^{-5} pg), then the oncogene will represent $1/10^6$ of the genome.

Table 2. Tests for oncogenicity of cellular DNA containing an activated *ras* gene and of cloned activated *ras* genes[a]

DNA			Animals tested	Route of injection	Tumour induction[b]	Reference
Type	Amount (µg)					
T24 genomic DNA	10	(H-*ras*)	Newborn rat	s.c.	0% (0/20)	10
Rat EJ-6 genomic DNA[c]	500	(H-*ras*)	Newborn hamster	s.c.	0% (0/8)	3
Chromatin from T24 cells	~1000		Rhesus monkey	Intramuscular, i.v., intracerebral (single or multiple doses)	In progress	(Food and Drug Administration, USA)
Cloned activated	2-10	(K-*ras*)	Newborn rat	s.c.	0% (0/10)	10
H- or *K-ras* genes	2-10	(H-*ras*)	Newborn rat	s.c.	0% (0/14)	10
	50	(H-*ras*)	Newborn hamster	s.c.	0% (0/15)	3

[a]The observation periods in these tests did not exceed two years.

[b]Percentage of animals in which tumours developed; actual numbers are given in parentheses.

[c]Rat EJ-6 cells are a line of rat embryo cells transformed by the *H-ras* oncogene from EJ bladder carcinoma.

Quantification of heterogeneous contaminating DNA or oncogenic sequences
Residual cellular DNA in biological products is generally quantified by nucleic acid hybridization techniques. For these tests the following probes have been used:

(a) Highly repetitive species-specific sequences that occur in high copy numbers in the cellular genome, such as the *Alu* sequences in human DNA.
(b) The specific gene sequences of interest and known to be present in the cell, such as those of a virus or of an oncogene.

In addition it may be useful to have available a technique that can detect DNA in general, irrespective of its sequence.

References

1. ISRAËL, M.A. ET AL. Biological activity of polyoma viral DNA in mice and hamsters. *Journal of virology*, **29**: 990-996 (1979).
2. BOUCHARD, L. ET AL. Tumorigenic activity of polyoma virus and SV40 virus DNAs in newborn rodents. *Virology*, **135**: 53-64 (1984).
3. MUFSON, R.A. & GESNER, T. Lack of tumorigenicity of cellular DNA and oncogene DNA in newborn hamsters. *In vitro monograph*, **6**: 168 (1985).
4. SOL, C.J.A. & VAN DERNOORDAA, J. Oncogenicity of SV40 DNA in the Syrian hamster. *Journal of general virology*, **37**: 635-638 (1977).
5. BURNETT, J.P. & HARRINGTON, J.A. Simian and adenovirus SA, DNA: chemical, physical and biological studies. *Proceedings of the National Academy of Sciences of the United States of America*, **60**: 1023-1029 (1968).
6. BURNETT, J. P. ET AL. Retention of tumour-inducing capacity by adenovirus DNA after cleavage by restriction endonucleases. *Nature (London)*, **254**: 158-159 (1975).
7. JOCHEMSEN, H. *Studies on the transforming genes and their products of human adenovirus types 12 and 5.* University of Leiden, 1979 (PhD thesis).
8. FUNG Y-K. T. ET AL. Tumor induction by direct infection of cloned v-src DNA into chicken. *Proceedings of the National Academy of Sciences of the United States of America*, **80**: 353-357 (1983).
9. WILL, H. ET AL. Cloned HBV DNA causes hepatitis in chimpanzees. *Nature (London)*, **299**: 740-742 (1980).
10. LEVINSON, A. D. ET AL. Tumorigenic potential of DNA derived from mammalian cell lines. *In vitro monograph*, **6**: 161-165 (1985).
11. MOYER, M.P. & MOYER, R.C. The effect of oncogenes inoculated into hamsters. *In vitro monograph*, **6**: 140-147 (1985).

Annex 2

TRANSFORMING PROTEINS

Proteins encoded by cellular transforming genes (*c-onc*) or by their mutated versions can induce the proliferation of many different cell types. Some onc proteins exert their biological effects from outside the cell, but most are active only intracellularly. *Onc* genes were first detected as the transforming genes of acute transforming retroviruses (*1, 2*). The sequences of the cellular homologues (proto-*onc*) of retroviral oncogenes have been highly conserved during evolution, and are similar in invertebrates and human beings. Many proto-*onc* appear to possess little or no oncogenic potential. However, when inserted into a retrovirus vector, at least some inappropriately regulated proto-*onc* or genes carrying structural mutations have an oncogenic potential similar to that of the viral oncogenes.

The profound alterations in cell growth that these genes can induce have prompted the question of what might be the tumorigenic risk of contamination by oncogenes or their protein products when cell-derived materials are administered to human beings. What, then, are the theoretical and actual hazards that might arise from oncogene-encoded proteins? The potential problems raised by contamination with normal or mutated proto-*onc* themselves have been discussed previously (*3*). In this discussion, the term oncogenes will include any gene whose encoded product can directly stimulate cell proliferation. Under this broad definition it will be possible to discuss growth factors that could not be included if a more stringent definition were used (*4, 5*).

Theoretical risks from DNA versus protein

In addition to obvious chemical differences between nucleic acids and proteins, at least two important qualitative distinctions between genes and their products are directly relevant to their oncogenic potential. Malignant transformation is believed to represent a more or less irreversible change in the proliferative state of the cell. Inadvertent administration of oncogene DNA could in theory cause changes that mimic this process, if the gene were taken up by the cell, since stable acquisition of the DNA could lead to the constitutive expression of its encoded onc protein. The introduction of the transforming gene may therefore cause an irreversible genotypic change in the cell, which results in the continuous exposure of the cell to the onc protein in a biologically relevant site, whether it is extracellular or intracellular. By contrast, the inadvertent administration of protein is theoretically less hazardous because, unlike DNA, it does not have the capacity for self-renewal. Its effects are therefore likely to be limited strictly by the amount that is administered, and will last only as long as the protein remains biologically active.

A second significant difference is that administered onc proteins can be active only if they exert their biological effects via the surface of the cell, since the uptake by the cell of onc proteins administered by standard techniques is inherently so inefficient that they could not, even theoretically, achieve high enough intracellular levels to be biologically active (*6*). This means that most onc proteins would not be potentially oncogenic as contaminants, since most of them are active as intracellular (as opposed to extracellular) proteins. Therefore only those onc proteins that are active extracellularly will be considered further. However, the general principles that follow should still be relevant to considering the potential hazards of onc proteins that would act intracellularly if administered under conditions favouring their cellular uptake in a biologically active form, for example within lipid-bound vesicles.

Growth factors

The onc proteins that can induce proliferation from outside the cell are usually referred to generically as peptide growth factors (*7*). The platelet-derived growth factor is the prototype for this group of molecules that can deliver an extracellular, growth-stimulatory signal (*8, 9*). This group also includes epidermal growth factor, the closely related transforming growth factor alpha, transforming growth factor beta, fibroblast growth factor, insulin, somatomedins (insulin-like growth factors), transferrin, and gastrin-releasing peptide (bombesin-like). Growth factors may also include molecules that initially were not believed to possess growth-promoting properties, such as tumour necrosis factor (*10*). Growth factors can be assayed quantitatively either biologically using susceptible cells or by specific radioimmunoassays.

The simian sarcoma virus is the only transforming retrovirus whose oncogene, v-*sis*, is known to encode a growth factor; v-*sis* is derived from one of the two genes for platelet-derived growth factor. In contrast to the intracellularly active protein products of several cellular transforming genes, no activated versions of platelet-derived growth factor or other growth factors with increased growth-promoting properties (compared with the normal versions) have been described. In particular, v-*sis* does not have a higher transforming activity than its cellular proto-*onc*. Growth factors such as platelet-derived growth factor may, under physiological circumstances (as in wound healing or clot formation following wound healing), be present in sufficiently high concentrations locally to induce cell growth in response to injury. It is important to note that the effects of growth factors are reversible and that these factors are not known to be specifically mutagenic. Inadvertent administration of growth factors as contaminants would represent exposure to products that are expressed by normal cells under certain physiological conditions. Tumours might arise only if "pharmacological" (i.e., non-physiological) doses were given repeatedly, and such tumours would presumably stop growing when the growth factor was no longer administered.

Each growth factor transmits its growth signal via specific receptors located on the cell surface. Therefore, a growth factor can act only on those cells that display the appropriate receptor. Receptors for many growth factors are found in only a subset of cells, although receptors for transforming growth factor beta are virtually ubiquitous in their distribution; this wide distribution among many different cell types may be related to the many apparent functions of transforming growth factor beta, which apparently stimulates cell growth under some circumstances and inhibits it under others.

As noted above, growth factors are synthesized physiologically by normal cells. Their synthesis may be elevated in various pathological states, and many tumour cells constitutively synthesize large amounts of various growth factors (*11*). The expression of certain oncogenes whose proteins act intracellularly has been shown to stimulate the production of secreted growth factors. The inappropriately elevated secretion of growth factors by tumour cells that display cell-surface receptors for those factors forms the basis for the hypothesis that some tumours persist via an autocrine mechanism. Increased levels of growth factor receptors, as reported for some tumours, may render the tumour cells more sensitive to growth induction by growth factors via paracrine or autocrine mechanisms (*12*).

Oncogenic potential of growth factors

It does not appear that growth factors, either singly or in combination, can by themselves induce the fully tumorigenic phenotype in primary cells. However, a combination of growth factors can induce established non-neoplastic rodent cells to become fully anchorage independent for growth, a phenotype that is often highly correlated in fibroblasts with tumorigenicity (*13*). Single growth factors can induce anchorage independence and tumorigenic growth of some established cells (*14*).

Oncogenic effects of growth factors have been demonstrated *in vivo*, but only under exceptional circumstances. In a strain of mice with a high incidence of mammary tumours whose continued growth depends in part on epidermal growth factor, surgical removal of the salivary glands, which are the major source of endogenous epidermal growth factor, led to a significant reduction in the incidence of mammary tumours (from 62% to 12%; *15*). Administration of "pharmacological" amounts of epidermal growth factor (5 μg per animal) every other day partially restored (to 33%) the normal tumour incidence. The results confirm that in this strain of mice tumour formation depends in part on epidermal growth factor. In another experiment, co-inoculation into a nude mouse of estrogen-independent and estrogen-dependent cells from mouse mammary tumours rendered the latter cells hormone independent, probably because of the production of autostimulatory growth factors (*16*).

In the above examples, growth factors were tumorigenic under circumstances in which the animals presumably contained markedly abnormal cells; the addition of growth factors probably represented only one of several important changes that resulted in tumorigenesis.

The oncogenic potential of the simian sarcoma virus appears to be extremely limited, despite the constitutive expression of v-*sis* in infected cells; when the virus does induce tumours, they are slowly growing fibroblastic masses that do not appear to be locally invasive, nor do they metastasize (*17*).

Quantitative considerations

Although some growth factors may be bound specifically to serum proteins, most have an extremely short half-life in the blood (2 min for platelet-derived growth factor). A short half-life markedly limits potential *in vivo* effects. *In vivo* administration of transforming growth factor alpha to newborn mice (which are highly susceptible to it) at daily subcutaneous doses of greater than 0.3 μg/g of body weight accelerated incisor eruption and eyelid opening (known biological effects of epidermal growth factor), but doses greater than 0.3 μg/g of body weight did not have any growth-promoting effects

(*18*). Usually 0.1-20 ng/ml of a growth factor are required for significant growth-promoting effects on responsive cells *in vitro*, although some factors, such as fibroblast growth factor, may be active at concentrations of several picograms per millilitre.

Even those cells that secrete large quantities of growth factors produce concentrations of no more than 50-75 ng/ml of a given growth factor in the culture medium. Because of dilution in body fluids, it would be almost impossible for administration of unconcentrated supernatant material to have any detectable oncogenic activity *in vivo*. Potential problems might arise, however, if growth factors were concentrated along with the product of interest. It seems reasonable to conclude provisionally that doses of growth factor at the level of 10 μg/kg of body weight daily would be without oncogenic effect, even in susceptible individuals. The murine mammary tumour system (in conjunction with salivary gland removal) might be a fruitful one to study further with different doses of epidermal growth factor or transforming growth factor alpha to obtain more definitive information on the risks of *in vivo* administration to a highly susceptible animal.

Summary and conclusions

The theoretical risk from oncogene-encoded proteins is limited to growth factors, which may be secreted by cells in which biological products might be manufactured. Because these peptides do not replicate, their effect is finite. In addition, their effects are reversible. Growth factors do not ordinarily appear to be oncogenic. Even under circumstances in which they contribute to oncogenesis, repeated administration of high concentrations of growth factors (several micrograms per kilogram) appears to be required for them to serve as cofactors in the carcinogenic process, and the resulting tumours appear to remain dependent upon the continued presence of the growth factor for their growth.

References

1. BISHOP, J.M. Viral oncogenes. *Cell*, **42**: 23 (1985).
2. WEINBERG, R.A. The action of oncogenes in the cytoplasm and nucleus. *Science*, **230**: 770 (1985).
3. Lowy, D. R. Potential hazards from contaminating DNA that contains oncogenes. *In vitro monograph*, **6**: 36 (1985).
4. CARPENTER, G. & COHEN, S. Epidermal growth factor. *Annual review of biochemistry*, **48**: 194 (1979).
5. ROBERTS, A.B. & SPORN, M.B. Transforming growth factors. *Cancer surveys*, **4**: 683 (1985).
6. BAR-SAGI, D. & FERAMISCO, J.R. Induction of membrane ruffling and fluid-phase pinocytosis in quiescent fibroblasts by *ras* proteins. *Science*, **233**: 1061 (1986).
7. SPORN, M.B. & ROBERTS, A.B. Peptide growth factors and inflammation, tissue repair, and cancer. *Journal of clinical investigation*, **78**: 329 (1986).
8. DEUEL, T.F. ET AL. Platelet derived growth factor: roles in normal and v-sis transformed cells. *Cancer surveys*, **4**: 633 (1985).

9. ROSS, R. ET AL. The biology of platelet derived growth factor. *Cell*, **46**: 155 (1986).
10. KOHASE, M. ET AL. Induction of Beta$_2$-interferon by tumor necrosis factor: a homeostatic mechanism in the control of cell proliferation. *Cell*, **45**: 659-666 (1986).
11. DICKSON, R.B. ET AL. Activation of growth factor secretion in tumorigenic states of breast cancer induced by 17β-estradiol or v-*ras*H oncogene. *Proceedings of the National Academy of Sciences of the United States of America* (in press).
12. ULRICH, A. ET AL. Human epidermal growth factor receptor cDNA sequence and aberrant expression of the amplified gene in A431 epidermoid carcinoma cells. *Nature (London)*, **309**: 418 (1984).
13. ASSOIAN, R.K. ET AL. Cellular transformation by coordinated action of three peptide growth factors from human platelets. *Nature (London)*, **309**: 804 (1984).
14. ROSENTHAL, A. ET AL. Expression in rat fibroblasts of a human transforming growth factor-alpha cDNA results in transformation. *Cell*, **46**: 301 (1986).
15. KURACHI, H. ET AL. Evidence for the involvement of the submandibular gland epidermal growth factor in mouse mammary tumorigenesis. *Proceedings of the National Academy of Sciences of the United States of America*, **82**: 5940 (1985).
16. DANIELPOUR, D. & SIRBASKU, D.A. New perspectives in hormone-dependent (responsive) and autonomous mammary tumour growth: role of autostimulatory growth factors. *In vitro*, **20**: 975 (1984).
17. DEINHARDT, F. Biology of primate retroviruses. In: Klein, G. ed. *Viral oncology*. New York, Raven Press, 1980, pp. 357-398.
18. TAM, J.P. Physiological effects of transforming growth factor in the newborn mouse. *Science*, **229**: 673 (1985).

This report contains the collective views of an international group of experts and does not necessarily represent the decisions or the stated policy of the World Health Organization

WHO Expert Committee on Biological Standardization

Thirty-sixth Report

World Health Organization
Technical Report Series (TRS)
745

World Health Organization, Geneva **1987**

WHO EXPERT COMMITTEE ON BIOLOGICAL STANDARDIZATION

Geneva, 12-18 November 1985

Members

Dr D.R. Bangham, Head, Division of Hormones, National Institute for Biological Standards and Control, London, England

Dr C. Guthrie, Operations Director, Commonwealth Serum Laboratories, Parkville, Victoria, Australia

Professor T.B. Jabloknva, Head, Laboratory of BCG and Tuberculin, Tarasevic State Research Institute for the Standardization and Control of Medical Biological Preparations, USSR Ministry of Health, Moscow, USSR *(Vice-Chairman)*

Dr H.W. Krijnen, Director, Central Laboratory of the Netherlands Red Cross Blood Transfusion Service, Amsterdam, The Netherlands *(Chairman)*

Mr J. Lyng, Head, Laboratory for Biological Standardization, State Serum Institute, Copenhagen, Denmark

Dr H. Mirchamsy, Associate Director, Razi State Institute of Serum and Vaccine Production, Tehran, Islamic Republic of Iran

Dr R. Murata, Honorary Member, National Institute of Health, Tokyo, Japan

Dr M.S. Nasution, Director, Perusahaan Umum "Bio Farma", Bandung, Indonesia

Dr W.W. Wright, Senior Scientist, Drug Standards Division, The United States Pharmacopeia, The National Formulary, Rockville, MD, USA *(Rapporteur)*

Dr Xiang Jian-zhi, Head, Division of Science and Technology, Shanghai Institute of Biological Products, Shanghai, China

Secretariat

Dr D.H. Calam Head, Division of Antibiotics and Chemistry. National Institute for Biological Standards and Control, London, England *(Temporary Adviser)*

Dr V. Grachev, Scientist, Biologicals, WHO, Geneva, Switzerland

Dr C. Hardegree, Director, Division of Bacterial Products, Office of Biologics Research and Review, Center for Drugs and Biologics, Food and Drug Administration Bethesda, MD, USA *(Temporary Adviser)*

Professor W. Hennessen, Berne, Switzerland *(Temporary Adviser)*

Dr F.H. Meskal, Director, National Research Institute of Health, Addis Ababa, Ethiopia *(Temporary Adviser)*

Dr J.C. Petricciani, Chief, Biologicals, WHO, Geneva, Switzerland *(Co-Secretary)*

Dr P. Sizaret, Scientist, Biologicals, WHO, Geneva Switzerland *(Co-Secretary)*

Dr D.P. Thomas, Head, Division of Blood Products, National Institute for Biological Standards and Control, London, England *(Temporary Adviser)*

Thirty-sixth Report

The WHO Expert Committee on Biological Standardization met in Geneva from 12 to 18 November 1985. The meeting was opened on behalf of the Director-General by Dr Lu Rushan, Assistant Director-General.

REQUIREMENTS FOR BIOLOGICAL SUBSTANCES

37. Requirements for Continuous Cell Lines

Requirements for continuous cell lines employed as substrates were established in 1982 as part of the *Requirements for Poliomyelitis Vaccine (Inactivated)* (WHO Technical Report Series, No. 673, 1982, p. 33). The Committee noted that more experience had been gained in the use of continuous cell lines as substrates for the preparation of biological substances by recombinant DNA techniques (WHO/BS/85.1465 Rev. 2). The Committee agreed that there was a need for general requirements for the characterization of continuous cell lines for use as substrates for the preparation of biological substances.

The Committee noted also that such requirements had been formulated and distributed for comment. After making some minor amendments the Committee adopted the *Requirements for Continuous Cell Lines,* and agreed that they should be annexed to this report (Annex 3). The Committee agreed also that when requirements for individual products are developed or revised, details of quality control procedures for cell cultures should be written taking these requirements into consideration.

The Committee agreed further that these requirements should supersede those in Part D of the *Requirements for Poliomyelitis Vaccine* (Inactivated) (WHO Technical Report Series, No. 673, 1982, pp. 70-76) (see Annex 4 of this report).

Annex 3
REQUIREMENTS FOR CONTINUOUS CELL LINES USED FOR BIOLOGICALS PRODUCTION
(Requirements for Biological Substances No. 37)

INTRODUCTION

Since WHO requirements for continuous cell lines used as substrates in the production of inactivated virus vaccines were established in 1983 as part of *Requirements for Poliomyelitis Vaccine (Inactivated)* (*1*), there have been several technical advances concerning the characterization of continuous cell lines. In addition, more experience has been gained in the use of continuous cell lines for the preparation of inactivated vaccines and as substrates for the preparation of biological substances by recombinant DNA techniques. Consequently, there is a need to provide general requirements for the characterization and use of continuous cell lines for the preparation of inactivated viral vaccines and other biological substances for use in man. Specific requirements for individual products will be developed or revised as appropriate and it will be particularly important to consult WHO requirements for individual products for details on quality control procedures related to cell cultures.

The Requirements presented in this document are intended to supersede Part D of *Requirements for Poliomyelitis Vaccine (Inactivated)* (*1*).

GENERAL CONSIDERATIONS

Several types of cell have been employed as substrates for the preparation of inactivated viral vaccines. These include: (*a*) cells derived directly by primary culture (or by a low number of passages) of normal animal tissues (e.g., monkey kidney and rabbit kidney cells); (*b*) diploid cell lines of human and monkey origin (e.g., WI-38, MRCS and FRhL-2 cells) which have a finite capacity for serial propagation; and (*c*) continuous cell lines (e.g., VERO and CHO cells). Diploid cell lines have certain disadvantages for the production of some biologicals in that their growth characteristics make them difficult to adapt to large-scale industrial production. Moreover, they have a finite life span and may produce relatively low yields of some vaccine viruses or other biological products.

Many current approaches to the development of inactivated viral vaccines employing modifications of conventional methods or new biotechniques, including recombinant DNA methods and controlled gene expression, involve the use of continuous cell lines. The advantage of such cell lines is that they grow relatively rapidly, providing high yields of monolayer, or, in some cases, suspension cultures.

Continuous cell lines may have biochemical, biological, and genetic abnormalities. In particular, they may produce transforming proteins and may contain viruses and potentially oncogenic DNA. In some cases, the cells may cause tumours when inoculated into animals. It is therefore important to ensure that biological products derived from continuous cell lines are highly purified. Generally, the purification procedures will result in the extensive removal of cellular DNA, other cellular components, and adventitious agents. Procedures that extensively degrade or denature DNA might also be considered.

The data required for the characterization of any continuous cell line to be used for production of biologicals include: (*a*) information on its origin, derivation and passage history; (*b*) information on its growth and morphological characteristics; (*c*) results of tests for adventitious agents; (*d*) distinguishing features, such as biochemical, immunological and cytogenetic patterns which allow the cells to be clearly recognized among other cell lines; and (*e*) results of tests for tumorigenicity, if this information is not already known.

Special considerations for control will apply to products derived from cells that contain known viral genomes (e.g., Namalva cells) or from cells modified by recombinant DNA technology.

It is important to establish a complete characterization of the continuous cell line so that appropriate tests for purity of the final product can be included. For example, if a cell line contains an endogenous virus, tests to ensure the absence of biologically active virus could be incorporated as one of the requirements for products derived from that cell line.

There has been considerable discussion internationally on general criteria for the acceptability of products (e.g., hormones, blood products, inactivated viral vaccines) prepared from transformed mammalian cells. A consensus has emerged on the desirability of achieving a high degree of purification of the product involving extensive removal or destruction of DNA of cell substrate origin. Manufacturers considering the use of transformed cells should be aware of the need to develop and authenticate efficient methods for the purification of the product and to establish sensitive methods for detection of cellular DNA as an essential element of any product development programme.

If transformed cells are being considered for the preparation of experimental live vaccines, careful consideration must be given to factors such as the possible incorporation of cellular DNA into the virions; considerable discussion will be required to assess the acceptability of such products.

Specific requirements for purity as well as other quality control procedures will be incorporated in WHO requirements for the individual products.

Each of the following sections constitutes a recommendation. The parts of each section that are printed in normal type have been written in the form of requirements so that, if a health administration so desires, these parts as they appear may be included in definitive national requirements. The parts of each section that are printed in small type are comments and recommendations for guidance.

Should individual countries wish to adopt these requirements as the basis of their national regulations concerning the use of continuous cell lines in the preparation of biological products, it is recommended that a clause should be included permitting modifications of manufacturing requirements on the condition that it can be demonstrated, to the satisfaction of the national control authority, that such modified requirements ensure that the degree of characterization of the continuous cell line is at least equal to those provided by the requirements formulated below. The World Health Organization should then be informed of the action taken.

The terms "national control authority" and "national control laboratory", as used in these requirements, always refer to the country in which the vaccine or other biological is manufactured.

PART A. MANUFACTURING REQUIREMENTS

1. DEFINITIONS

Cell seed: A quantity of cells of human or animal origin stored frozen at -70° C or below in aliquots of uniform composition, one or more of which would be used for the production of a manufacturer's working cell bank.

Manufacturer's working cell bank (MWCB): A quantity of cells derived from one or more ampoules of the cell seed and of uniform composition, stored frozen at -70° C or below in aliquots, one or more of which would be used for production purposes.

In normal practice a cell seed is expanded by serial subculture up to a passage number (or population doubling, as appropriate) selected by the manufacturer, at which point the cells are combined into one pool and preserved cryogenically to form the MWCB. One or more of the ampoules from such a pool would be used for production purposes.

Production cell culture: A collection of cell cultures being used for biological production that have been derived from one or more ampoules of the MWCB.

Adventitious agents: Contaminating microorganisms of the cell line, including bacteria, fungi, mycoplasmas, and endogenous and exogenous viruses.

2. GENERAL MANUFACTURING REQUIREMENTS

The general requirements contained in the revised *Requirements for Biological Substances No.1 (General Requirements for Manufacturing Establishments and Control Laboratories)* (*2,* page 11) shall apply but with the provision that during the production of a given biological no cell cultures other than those approved by the national control authority for the production of the appropriate biological shall be introduced or handled in the production area.

3. IDENTIFICATION AND CHARACTERISTICS OF CELL LINE

3.1 Cell seed and manufacturer's working cell bank (MWCB)

The use of continuous cell lines for the manufacture of biological products shall be based on the cell seed system. A continuous cell line shall be subcultured to a point at which it is convenient to prepare a cell seed. The passage level, or population doubling, of the cell seed should be as low as possible. The cell seed used for the production of biologicals shall be that approved by and registered with the national control authority.

The continuous cell line from which the cell seed and the MWCB have been derived shall have been characterized with respect to genealogy and growth characteristics at passage levels (or population doublings, as appropriate) equivalent to those of the cell seed, MWCB, and the cell cultures used for production. In addition, information on specific immunological markers and storage conditions will be of value. Also the cell line shall have been tested in animals, eggs, and cell culture for detectable adventitious agents. Virus susceptibility and capacity for interferon production may be other useful markers for cell identification. In addition, cytogenetic data may be useful for cell identification as well as for assessing the genetic stability of the cell line used for production purposes. However, a complete cytogenetic characterization of the cell line is not recommended because of the limited usefulness of such information.

These data shall be made available to the national control authority.

3.2 Tests for adventitious agents

Tests in the following sections are intended to identify any endogenous or exogenous agents that may be present in the cell line. Special attention should be given to tests for agents known to be present in a latent state in the species from which the cells were derived (e.g., SV-40 virus in rhesus monkeys).

3.2.1 *Tests in animals and eggs*

The tests in animals and eggs for adventitious agents that can induce pathology shall include the inoculation by the intramuscular route of each of the following groups of animals with cells from the MWCB propagated to or beyond the maximum passage level (or population doubling, as appropriate) used for production, using at least 10^7 viable cells divided equally among the animals in each group:

- 2 litters of suckling mice, comprising at least 10 animals, less than 24 h old;
- 10 adult mice;
- 5 guinea-pigs; and
- 5 rabbits.

> The test in rabbits for the presence of B virus in the cell line may be replaced by a test in rabbit kidney cell cultures.

At least 10^6 viable cells shall also be injected into the allantoic cavity, the chorioallantoic membrane, and yolk sac of each of 10 embryonated chicken eggs 9-11 days old.

Also at least 10^6 viable cells shall be injected intracerebrally into each of 10 adult mice to test for the presence of lymphocytic choriomeningitis virus.

The animals shall be observed for at least 4 weeks and the embryonated chicken eggs shall be examined after not less than 5 days of incubation. Any animals that are sick or show any abnormality shall be investigated to establish the cause of illness. The allantoic fluids shall be tested with guinea-pig and chick or other avian red cells for the presence of haemagglutinins.

The cells are suitable for production if at least 80% of the animals or eggs inoculated with the cells remain healthy and survive the observation period and none of the animals or eggs shows evidence of the presence in the cell cultures of any adventitious agent.

3.2.2 *Sterility tests*
Tests for bacteria, fungi and mycoplasmas shall be performed according to the *General Requirements for the Sterility of Biological Substances* (*3*).

3.2.3 *Morphological tests*
The cells from the MWCB propagated to the maximum passage level used for production, or a level beyond it, shall be examined and characterized by light and electron microscopy to establish their morphological features. Comparisons of these data with those derived from cells of a passage level close to that of the MWCB will be of value in assessing the stability of the cell line and in determining the presence of potential microbial contaminants.

The cells, propagated to at least 10 population doublings beyond the maximum level used for production, shall also be tested for the presence of endogenous viruses. Suitable methods include induction with agents such as idoxuridine (IUDR) and the examination of electron micrographs for virus particles.

3.2.4 *Tests on cell cultures*
(1) Intact cells should be cultivated along with a range of other cell systems, including cells derived from man, and examined for morphological changes. Disrupted cells should be inoculated onto a similar set of cell culture systems and examined as above.
(2) Cell culture fluids should be examined for the presence of cytopathic agents in cell cultures of various species, including human cells, and at the end of the observation period the cell culture fluids should be tested for haemagglutinating viruses.

3.2.5 *Tests for retroviruses*
In addition to the morphological tests described in section 3.2.3, assays for viral reverse transcriptase should be performed on cells that have been induced with chemicals such as idoxuridine (IUDR).

3.2.6 *Other tests*
Additional studies may be required to characterize and identify the adventitious agents found in any of the tests required in section 3.2.

3.3 Tests for tumorigenicity
3.3.1 In vivo *tests*
Cells from the MWCB, propagated to at least 10 population doublings beyond the maximum passage level used for production, shall be examined for tumorigenicity in a test approved by the national control authority. The test should involve a comparison between the continuous cell line and a suitable positive control reference preparation (e.g., HeLa or Hep 2 or FL cells). Information concerning the sources of suitable positive

control reference cells may be obtained from Chief, Biologicals, World Health Organization, 1211 Geneva 27, Switzerland. A negative control is not essential but desirable. For that purpose nontumorigenic diploid cell lines such as WI-38 or MRC-5 may be used.

Animal systems that have been shown to be suitable for this test include:
(*a*) athymic mice (Nu/Nu genotype); or
(*b*) newborn mice, rats or hamsters that have been treated with antithymocyte serum or globulin; or
(*c*) thymectomized and irradiated mice that have been reconstituted (T-, B+) with bone marrow from healthy mice.

Whichever animal system is selected, the cell line and the reference cells are injected into separate groups of 10 animals each. In both cases, the inoculum for each animal is 10^7 cells suspended in a volume of 0.2 ml, and the injection may be by either the intramuscular or subcutaneous route. In the case of newborn animals (system *b*) the animals are treated with 0.1 ml of antithymocyte serum or globulin on days 0, 2, 7, and 14 after birth.

A potent serum or globulin is one that suppresses the immune mechanisms of the growing animals to the extent that the subsequent inoculum of 10^7 positive reference cells regularly produces tumours and metastases. Animals in all tests are observed for 21 days for evidence of nodule formation at the site of injection. If nodules appear, they are measured in two perpendicular dimensions, the measurements being recorded regularly to determine whether there is progressive growth of the nodule.

At the end of the observation period all animals, including the reference group(s), are sacrificed and examined for gross and microscopic evidence of the proliferation of inoculated cells at the site of injection and in other organs (e.g., lymph nodes, lungs, kidneys, and liver).

In a valid test, progressively growing tumours should be produced in at least 9 of 10 animals injected with the positive reference cells.

> If the continuous cell line has already been documented to be tumorigenic, or if the class of cells to which it belongs (e.g., hybridomas) is tumorigenic, the national control authority may not require additional tumorigenicity tests.
>
> Tests for the oncogenic potential of cellular DNA and crude cell lysates in animal systems may also be useful in establishing the basic characteristics of the cells.

3.3.2 In vitro *tests*

Two *in vitro* tests that have been found to correlate well with tumorigenicity are: (*a*) colony formation in soft agar gels; and (*b*) production of invasive cell growth following inoculation onto organ cultures.[1] These may be used to characterize more fully the cell lines that show no evidence of tumorigenicity in animal tests (section 3.3.1) or when the results are equivocal.

In vitro tests may be considered sufficient to characterize the cells for tumorigenicity by some national control authorities.

A matter of particular concern is the possibility that the cells used in the production of biologicals may contain activated oncogenes. For that reason, assays of cell transformation with DNA derived from the cell line (at the passage level used for production) should be performed in order to determine whether or not activated oncogenes can be detected in the continuous cell line. The standard 3T3 assay system should be used for these tests; but additional tests should also be considered as new techniques are developed for the detection of a broader range of oncogenes.

3.4 Identity test

The MWCB shall be identified by a method approved by the national control authority. Methods for identity testing include biochemical (e.g., isoenzyme analyses), immunological (e.g., histocompatibility antigen assays), and cytogenetic marker tests.

[1]Information concerning details of these test procedures can be obtained from Chief, Biologicals, World Health Organization, 1211 Geneva 27, Switzerland.

AUTHORS

The *Requirements for Continuous Cell Lines used for Biologicals Production* were formulated by the following WHO consultants and staff members:

Dr F. Assaad, Director, Division of Communicable Diseases, WHO, Geneva, Switzerland
Dr J. Furesz, Director, Bureau of Biologics, Drug Directorate, Ottawa, Canada *(Consultant)*
Dr V. Grachev, Scientist, Biologicals, WHO, Geneva, Switzerland
Professor F. Horaud, Medical Virology, Pasteur Institute, Paris, France *(Consultant)*
Professor L.O. Kallings, Director, National Bacteriological Laboratory, Stockholm, Sweden *(Consultant)*
Dr V.R. Kalyanaraman, Director, Pasteur Institute of India, Coonoor, India *(Consultant)*
Dr B. Montagnon, Mérieux Institute, Charbonnières les Bains, France *(Consultant)*
Dr R. Netter, Director-General, National Laboratory of Health, Ministry of Social Affairs and National Solidarity, Paris, France *(Consultant)*
Dr J. Petricciani, Director, Blood and Blood Products, Office of Biologics, Center for Drugs and Biologics, Food and Drug Administration, Bethesda, MD, USA *(Consultant)*
Dr G.C. Schild, National Institute for Biological Standards and Control, London, England *(Consultant)*
Dr P. Sizaret, Scientist, Biologicals, WHO, Geneva, Switzerland
Dr H. J. van de Donk, State Institute for Public Health and Environmental Hygiene, Bilthoven, Netherlands *(Consultant)*

ACKNOWLEDGEMENTS

Acknowledgements are due to the following experts for their comments and advice:
Mr I. Davidson, Central Veterinary Laboratories, Weybridge, England
Dr B. Elisberg, Office of Biologics Research and Review, Center for Drugs and Biologics, Bethesda, MD, USA
Dr E. Esber, Office of Biologics Research and Review, Center for Drugs and Biologics, Bethesda, MD, USA
Dr C. Guthrie, Commonwealth Serum Laboratories, Parkville, Victoria, Australia
Dr P. Lemoine, Institute of Hygiene and Epidemiology, Brussels, Belgium
Dr G. Mann, London School of Hygiene and Tropical Medicine, London, England
Dr B.C. Meyer, Office of Biologics Research and Review, Center for Drugs and Biologics, Bethesda, MD, USA
Dr P. Noguchi, Office of Biologics Research and Review, Center for Drugs and Biologics, Bethesda, MD, USA
Professor Oon Chong Jin, National University of Singapore, Singapore General Hospital, Singapore
Dr P. Parkman, Ofice of Biologics Research and Review, Center for Drugs and Biologics, Bethesda, MD, USA
Dr L. Peyron, Mérieux Institute, Charbonnières les Bains, France
Dr J.C. Vincent-Falquet, Mérieux Institute, Charbonnières les Bains, France

Dr Yuan-Yuan Chu, Office of Biologics Research and Review, Center for Drugs and Biologics, Bethesda, MD, USA
Professor A.J. Zuckerman, London School of Hygiene and Tropical Medicine, London, England

REFERENCES

1. WHO Technical Report Series, No. 673, 1982.
2. WHO Technical Report Series, No. 323, 1966.
3. WHO Technical Report Series, No. 530, 1973.

This report contains the collective views of an international group of experts and does not necessarily represent the actions or the stated policy of the World Health Organization

WHO Expert Committee on Biological Standardization

Forty-seventh Report

World Health Organization
Technical Report Series (TRS)
878

World Health Organization, Geneva **1998**

WHO Expert Committee on Biological Standardization
Geneva, 7-11 October 1996

Members

Dr D. Calam, National Institute for Biological Standards and Control, Potters Bar, Herts., England (*Rapporteur*)
Dr S. Drozdov, Director, Institute of Poliomyelitis and Viral Encephalitides, Moscow, Russian Federation
Dr I.D. Gust, Research and Development Director, CSL Ltd, Parkville, Victoria, Australia
Dr J.G. Kreeftenberg, Bureau for Quality Assurance and Regulatory Affairs, National Institute for Public Health and Environmental Protection, Netherlands (*Vice-Chairman*)
Dr F.A. Ofosu, Department of Pathology, McMaster University, Hamilton, Ontario, Canada
Dr J.C. Petricciani, Vice-President, Genetics Institute, Cambridge, MA, USA (*Chairman*)
Mr Zhou Hai-jun, Director, National Institute for the Control of Pharmaceutical and Biological Products, Temple of Heaven, Beijing, China

Representatives of other organizations

Council of Europe
Mr J.-M. Spieser, European Department for the Quality of Medicines, Council of Europe, Strasbourg, France
International Association of Biological Standardization
Professor F. Horaud, Pasteur Institute, Paris, France
Dr G. Schild, Director, National Institute for Biological Standards and Control, Potters Bar, Herts., England
International Federation of Pharmaceutical Manufacturers Associations
Dr M. Duchene, Director, Quality Control, SmithKline Beecham Biologicals, Rixensart, Belgium
Dr B. Montagnon, Head, Regulatory Affairs for Vaccines, Pasteur Mérieux Sera and Vaccines, Marcy l'Etoile, France
International Society on Thrombosis and Haemostasis
Dr A. Tripodi, Haemophilia and Thrombosis Centre, Milan, Italy

Secretariat

Dr W.G. van Aken, Central Laboratory of the Netherlands Red Cross Blood Tranfusion Service, Amsterdam, Netherlands (*Temporary Adviser*)
Dr F.S. Antezana, Assistant Director-General, World Health Organization, Geneva, Switzerland
Dr P. Corran, National Institute for Biological Standards and Control, Potters Bar, Herts., England (*Temporary Adviser*)
Dr V. Grachev, Deputy Director, Institute of Poliomyelitis and Viral Encephalitides, Moscow, Russian Federation (*Temporary Adviser*)
Dr E. Griffiths, Chief, Biologicals, World Health Organization, Geneva, Switzerland (*Secretary*)
Dr G. Hansen, Statens Seruminstitut, Copenhagen, Denmark (*Temporary Adviser*)
Dr M.C. Hardegree, Director, Office of Vaccine Research and Review, Center for Biologics Evaluation and Research, Food and Drug Administration, Rockville, MD, USA (*Temporary Adviser*)
Dr K. Haslov, Statens Seruminstitut, Copenhagen, Denmark (*Temporary Adviser*)
Dr J.E. Idänpään-Heikkilä, Director, Division of Drug Management and Policies, World Health Organization, Geneva, Switzerland
Dr A.M. Padilla, Scientist, Biologicals, World Health Organization, Geneva, Switzerland
Dr J. Robertson, National Institute for Biological Standards and Control, Potters Bar, Herts., England (*Temporary Adviser*)
Dr D. Thomas, Witney, Oxford, England (*Temporary Adviser*)
Dr K. Zoon, Director, Center for Biologics Evaluation and Research, Food and Drug Administration, Rockville, MD, USA (*Temporary Adviser*)

Requirements and guidelines for biological substances

Requirements for cell substrates used for production of biologicals

The Committee noted a draft of proposed requirements for the use of animal cells as *in vitro* substrates for the production of biologicals (BS/95.1792 Rev.5) that had been revised following extensive discussion and comment on earlier drafts by a wide group of experts, in particular at a meeting held during October 1996. The Committee was informed that the proposed requirements represented a major departure from previous requirements published by WHO in that residual cellular DNA was no longer regarded as a significant risk factor requiring removal to extremely low levels. The Committee was further informed that certain requirements for well established diploid cell lines, such as MRC-S and WI-38, had been relaxed. After making some additional modifications, the Committee adopted the text as Requirements for the Use of Animal Cells as *in vitro* Substrates for the Production of Biologicals, agreed that it should be annexed to its report (Annex 1) and noted that these requirements replaced the Requirements for Continuous Cell Lines Used for Biologicals Production (Requirements for Biological Substances, No. 37).

Annex 1

Requirements for the use of animal cells as *in vitro* substrates for the production of biologicals

(Requirements for Biological Substances No. 50)

Introduction

Historically, the major concerns regarding the quality of biological products produced in animal cells have been related to the possible presence of adventitious contaminants and, in some cases, to the properties of the cells themselves. There are additional concerns regarding the quality of products prepared using recombinant DNA technology in relation to the expression construct contained in the cell substrates. It is well established that the properties of cell substrates and events linked to growth can affect the quality of the resultant biological products and, furthermore, that effective quality control of these products requires appropriate controls on all aspects of the handling of cell substrates.

General considerations

Types of animal cell substrates

Primary cell substrates

Primary cells obtained directly from the trypsinized tissues of normal animals have played a prominent role in the development of virology as a science, and of immunology in particular. Cultures of primary cells from different sources have been in worldwide use for the production of live and inactivated viral vaccines for human use for more than 40 years, and experience has indicated that these products are safe and effective.

Major successes in the control of viral diseases, such as poliomyelitis, measles, mumps and rubella, were made possible through the wide use of vaccines prepared in primary cell cultures, including those from chicken embryos and the kidneys of monkeys, dogs, rabbits and hamsters, as well as other tissues. Cultures of monkey kidney cells have been used for the production of inactivated and oral poliomyelitis vaccines for more than 40 years, and the same cell system continues to be used for the production of both vaccines.

Primary cell cultures have the following advantages: they are comparatively easy to prepare using simple media and bovine sera; and they possess a broad sensitivity to different viruses, some of which are cytopathogenic. In addition, primary cells can now be grown in bioreactors using the microcarrier method (*1*). However, where suitable alternative cell substrates are available, primary cell cultures are less likely to be used in the future for the following reasons: contamination by infectious agents, such as viruses, is a common problem; the quality and sensitivity of cultures obtained from different animals is variable; and it will become increasingly difficult to obtain cultures derived from nonhuman primates.

Primary cell cultures obtained from wild animals show a high frequency of viral contamination. For example, it is generally accepted that monkey-kidney cell cultures can be contaminated with one or more adventitious agents, including simian viruses. The number of viruses isolated and the frequency of isolation depend on many factors, including the method of isolation, test cell systems used, number of passages and duration of incubation and co-cultivation, and are directly proportional to the incubation period of the cultures. The frequency of contaminated cell cultures can be significantly reduced by careful screening of the source animals for the absence of antibodies to relevant viruses. The use of animals bred in a carefully controlled colony, especially those which are specific-pathogen free, is strongly recommended. The use of secondary or tertiary cells on which testing for adventitious agents can be performed will also reduce the frequency of contaminated production cell cultures.

Diploid cell substrates

The essential features of diploid cell lines of human (e.g. WI-38, MRC-5) or monkey (FRhL-2) origin are: they have a finite capacity fur serial propagation, which ends in senescence; and they are nontumorigenic and display diploid cytogenetic characteristics with a low frequency of chromosomal abnormalities of number and structure. Substantial experience over the past 25 years has been accumulated on the karyology of WI-38 and MRC-5 diploid cell lines, and ranges of expected frequencies of chromosomal abnormalities have been published (*2*). More sophisticated cytogenetic techniques (e.g. banding) have demonstrated subtle chromosomal abnormalities that were previously undetectable, thus making the previously established ranges of abnormalities obsolete. Recent studies have shown that subpopulations of human diploid cells with such abnormalities may appear and disappear over time, and that they are non-tumorigenic and undergo senescence.

The possibility of using human diploid cell substrates for the production of viral vaccines was demonstrated more than 35 years ago. The experience gained with oral poliomyelitis and other viral vaccines in successfully immunizing millions of children in many countries has clearly demonstrated the safety of vaccines produced on such substrates (*3*).

The main advantage of diploid cell lines in comparison to primary cells is that they can be well characterized and standardized, and production can be based on a cell bank system. In addition, unlike the continuous cell lines discussed below, they possess a finite life and are not tumorigenic. The cell bank system usually consists of cell banks of defined passage levels and may include a master cell bank and a working cell bank.

However, diploid cell lines have the following disadvantages: they are not easy to use in large-scale production, such as bioreactor technology employing the microcarrier method; in general, they need a more demanding growth medium than other cell substrates; and they usually need larger quantities of bovine serum (either fetal or donor calf) for their growth than do continuous cell lines.

Continuous-cell-line substrates

Continuous cell lines have the potential for an infinite life span and can usually be cultivated as attached cells or in suspension in a bioreactor. They have been derived by the following methods: (a) serial subcultivation of a primary cell culture of a human or animal tumour cell, such as HeLa or Namalva cells; (b) transformation of a normal cell having a finite life span with an oncogenic virus, for example, a B lymphocyte transformed by the Epstein-Barr virus; (c) serial subcultivation of a normal cell population generating a new cell population having an infinite life span; or (d) fusion between a myeloma cell and an antibody-producing B lymphocyte.

While cell transformation can occur spontaneously in various animal cells grown *in vitro* (continuous cell lines from African green monkey kidney cells (Vero), baby hamster kidney cells (BHK21) and Chinese hamster ovary cells (CHO) were established in this way), it has not been reported with human cells derived from normal tissues.

Hybridoma cells express monoclonal antibodies and hybridoma cell lines have generally been established from rodent hybridomas. Human hybridomas are obtained by the transformation of a B lymphocyte with Epstein-Barr virus, usually followed by subsequent fusion with a murine myeloma cell.

Continuous cell lines are now considered to be suitable substrates for the production of many biological medicinal substances and possess distinct advantages over primary and diploid cell substrates (*4*). A cell bank system similar to that used for diploid cell lines provides a means for the production of biologicals for an indefinite period based on well characterized and standardized cells. Continuous cell lines tend to be less demanding than diploid cell lines; as a rule they grow well using ordinary media and serum, and some do not require serum at all. They can also be used in microcarrier cultures and/or suspension cultures for large-scale production in bioreactors.

However, many continuous cell lines express endogenous viruses and are tumorigenic. Their theoretical disadvantages therefore include the risk of tumorigenicity associated with residual cellular DNA that may encode transforming proteins.

In 1986, a WHO Study Group considered a number of issues associated with the acceptability of new cell substrates for the production of biologicals (*5*) and concluded that, in general, continuous cell lines were acceptable for this purpose, but that differences in the nature and characteristics of the products and in manufacturing processes must be taken into account when making a decision on the use of a particular continuous cell line in the manufacture of a given product. WHO Requirements for Continuous Cell Lines used for Biologicals Production were published in 1987 (*6*).

In addition, the WHO Study Group recommended the establishment of well characterized cell lines that would be of value to national control authorities and manufacturers of biologicals. In following up this recommendation, WHO developed a WHO master cell bank for Vero cells, a continuous cell line established from the kidneys of African green monkeys. The reason for selecting this cell line was that it offered the immediate prospect of improving the quantity and quality of several vaccines being produced in other systems.

A master cell bank of Vero cells was donated to WHO, by a manufacturer, at the 134th passage. The maximum passage level recommended for production is 150. Studies of tumorigenicity in newborn rats suggest that cells in the passage range 134-150 are not tumorigenic. Collaborative studies in 10 laboratories with respect to sterility, adventitious agents, tumorigenicity, presence of reverse transcriptase and identity showed that the WHO Vero cell bank met the WHO Requirements for Continuous Cell Lines used for Biologicals Production (*6*).

The WHO master cell bank of Vero cells is stored at the European Collection of Animal Cell Cultures (ECACC), Porton Down, England and the American Type Culture Collection (ATCC), Rockville, MD, USA. Producers of biologicals and national control authorities can obtain cultures of these Vero cells (free of charge), as well as additional background information, from Biologicals, World Health Organization, 1211 Geneva 27, Switzerland (*7*).

Potential risks associated with biologicals produced in animal cells

The main potential risks associated with the use of biologicals produced in animal cells are directly related to contaminants from the cells, and they fall into three categories: viruses and other transmissible agents; cellular DNA; and growth-promoting proteins. A summary of the risk assessment for each follows. More comprehensive statements have been published elsewhere on the risks associated with contaminating DNA (*5*, *8-15*) and growth-promoting proteins (*5*).

Viruses and other transmissible agents

The 1986 WHO Study Group reviewed the potential risk to human recipients of products manufactured in cells containing viral agents. These may include complete viruses with known patterns of replication, such as simian virus 40 (SV40), virus particles such as type A retroviruses, which can be visualized by electron microscopy, and persisting viral genomes or parts of genomes, for example those of the hepatitis B and Epstein-Barr viruses. As described below, cells differ with respect to their potential for carrying viral agents pathogenic for human beings.

Primary monkey-kidney cells have been used to produce hundreds of millions of doses of poliomyelitis vaccines over the past 40 years, and although latent viruses, such as simian virus 40, were discovered in these cells, control measures were introduced to eliminate the risk associated with the manufacture of vaccines in cells containing those endogenous viruses. Additional controls may be needed as new viral agents and technologies are identified.

Human and nonhuman primate lymphocytes and macrophages may carry latent viruses, such as herpesvirus and retroviruses. Continuous lines of non-haematogenous cells from human and nonhuman primates may contain viruses or have viral genes integrated into their DNA. In either case, virus expression may occur under *in vitro* culture conditions.

Avian tissues and cells harbour exogenous and endogenous retroviruses, but there is no evidence for transmission of disease to humans from products prepared using these substrates. For example, large quantities of yellow fever, measles and live influenza vaccines have been produced for many years in eggs that contain avian leukosis viruses, but there is no evidence that these products have had any harmful effects in their long history of use for human immunization.

Rodents harbour exogenous and endogenous retroviruses. Lymphocytic choriomeningitis virus and haemorrhagic fever viruses from rodents have caused disease in humans by direct infection.

Human diploid fibroblasts have been used for vaccine production for over 30 years, and although concern was initially expressed about the possibility of the cells containing a latent human virus, no evidence for such an agent has been found, and vaccines produced from this class of cell have proved to be free from viral contaminants.

In light of the differing potential of the various types of cells mentioned above for transmitting viruses pathogenic in humans, different types of testing are appropriate for products manufactured using these cells.

When either diploid cell lines or continuous cell lines are used for production, a cell bank system is used and the cell bank is characterized as specified in the appropriate requirements published by WHO. Additional methods such as testing for viral sequences or other viral markers should also be considered. Efforts to identify viruses constitute an important part of the characterization of cell banks.

When cell lines of rodent or avian origin are examined for the presence of viruses, the major emphasis in risk assessment is placed on the results of studies in which transmission to target cells or animals is attempted. Risk to human recipients should not be assessed solely on ultrastructural evidence of the presence of viral agents in the cells.

The overall manufacturing process, including the selection and testing of cells and source materials, any purification procedures used and tests on intermediate or final products, has to be such as to ensure the absence of detectable infectious virus in the final product.

There may be as yet undiscovered microbial agents for which there is no current evidence or means of detection. As such agents become identified, it will be important to re-examine cell systems for their presence. Positive findings will have to be discussed with the national control authority.

Cellular DNA

Primary and diploid cells have been used successfully and safely for many years for the production of viral vaccines, and the residual cellular DNA deriving from these cells has not been (and is not) considered to pose any risk. Continuous cell lines have an infinite life span due to the deregulation of genes that control growth. The DNA deriving from such cell lines is therefore considered to have the potential to confer the capacity for unregulated cell growth, or tumorigenic activity, upon other cells.

The 1986 WHO Study Group advised on the levels of contaminating DNA deriving from continuous cell lines used in the production of biologicals for human use (*5*). Risk assessment based on an animal oncogene model suggested that *in vivo* exposure to one nanogram (ng) of cellular DNA, where 100 copies of an activated oncogene were present in the genome, would give rise to a transformational event once in 10^7 recipients (*13*). On the basis of this and other available evidence, the Study Group concluded that the risk associated with residual continuous-cell-line DNA in a product is negligible when the amount of such DNA is 100 picograms (pg) or less per parenteral dose. In determining this limit, the perceived problem was not the DNA itself but rather minimizing the presence of specific DNA sequences coding for activated oncogenes.

Additional calculations suggest that the risk of insertional mutagenesis that could lead to a neoplastic event is extremely small. In one recent report, it was predicted that a 10-ug dose of DN A would result in the inactivation of two independent tumour-suppressor genes, by insertional mutagenesis, within a single cell of a vaccine recipient in only one of 10^7 recipients (*9*). These very low calculated levels of risk are consistent with the limited human and animal experience to date (*10*, *16-18*).

Additional data published recently have shown that milligram amounts of DNA containing an activated oncogene from human tumour cells have not caused tumours in nonhuman primates during an evaluation period of 10 years (*16*). Also, human blood contains substantial amounts of DNA in plasma (75-450 ug per unit of blood) (*19*, *20*). Furthermore, contaminating DNA in a biological product generally occurs as small fragments unlikely to encode a functional gene.

The assessment of the safety of a product with respect to residual cellular DNA has to take into account: (a) the low levels of risk implied by the considerations described above; (b) the possible inactivation of any biological activity of contaminating DNA during processing; and (c) any reduction in the level of contaminating DNA during the purification process. A product may be considered safe on the basis of (b) and/or (c).

The current state of knowledge suggests that continuous-cell-line DNA can be considered as a cellular contaminant, rather than as a significant risk factor requiring removal to extremely low levels. On the basis of this reassessment, the Expert Committee concluded that levels of up to 10 ng per purified dose can now be considered acceptable. The purification process has to be validated by appropriate methods, including spiking studies, to demonstrate its capability to remove DNA to an acceptable level. In addition, batch-to-batch consistency needs to be shown for clinical trial batches and for three or more consecutive production batches. Subsequently, routine release testing for continuous-cell-line DNA in the final purified batch may not be needed. Any exceptions need to be agreed with the national control authority. For example, data suggest that beta-propiolactone, a viral inactivating agent, may also destroy the biological activity of DNA; use of this agent therefore provides an additional level of confidence even when the amount of DNA per parenteral dose may be substantial (*21*). Data should be obtained on the effects of such inactivating agents under specific manufacturing conditions so that firm conclusions on their DNA-inactivating potential for a given product can be drawn.

There may be instances where continuous-cell-line DNA is considered to pose a greater risk, e.g. where it could include infectious retroviral provirion sequences. Under these circumstances, acceptable limits should be set in consultation with the national control authority.

The new upper limit of 10 ng of residual DNA per dose does not apply to products derived from microbial, diploid or primary-cell-culture systems. The 1986 WHO Study Group stated that the risks for continuous-cell-line DNA should be considered negligible for preparations given orally: for such products, the principal requirement is the elimination of potentially contaminating viruses and toxic proteins. The upper limit of 10ng of residual continuous-cell-line DNA per dose therefore does not apply to a product given orally. Acceptable limits should be set in consultation with the national control authority.

Growth-promoting proteins

Growth factors may be secreted by cells used to produce biologicals, but the risks from these substances are limited, since their growth-promoting effects are usually transient and reversible, they do not replicate, and many of them are rapidly inactivated in vivo. In exceptional circumstances, growth factors can contribute to oncogenesis, but even in these cases, the tumours apparently remain dependent upon continued administration of the growth factor. Therefore, the presence of known growth-factor contaminants at ordinary concentrations does not constitute a serious risk in the preparation of biological products from animal cells.

Proteins prepared using continuous-cell-line substrates need to be purified to permit their safe clinical use. Analytical methods to assure the purity of each batch should be proposed and validated by the manufacturer. The purification process should also be validated to demonstrate its capability to remove host-cell proteins to an acceptable level. In addition, batch-to-batch consistency should be shown for clinical trial batches and for three or more consecutive production batches. Subsequently, routine release testing for host cell proteins in the final purified batch may not be needed.

Requirements published by WHO

The first requirements published by WHO for cell cultures used for the production of biologicals were formulated in 1959 for the production of inactivated poliomyelitis vaccine in primary cell substrates (*22*). They were revised in 1965 (*23*). The successful use of primary cell cultures derived from the kidneys of clinically healthy monkeys for the production of both inactivated and oral poliomyelitis vaccine (*24*) led to confidence in the use of other cell cultures for the production of various viral vaccines. Many types of cell culture are now widely used for the production not only of viral vaccines, but also of other biologicals, such as monoclonal antibodies and a wide range of biologicals prepared using recombinant DNA technology.

Taking into account the latest available data relating to cell substrates and after extensive consultation, especially at a WHO/International Association of Biological Standardization/Mérieux Foundation International Symposium on the Safety of Biological Products prepared from Mammalian Cell Culture held in Annecy, France, in September 1996, the WHO Expert Committee on Biological Standardization adopted the text of this Annex as requirements appropriate for the quality control of animal cells used as *in vitro* substrates for the production of biologicals. They supersede previous requirements describing procedures for the growth and quality control of cell substrates for the production of biologicals (*5*, *6*) and should be read in conjunction with the requirements published by WHO for individual products.

The following requirements concern the characterization and testing of continuous-cell-line and diploid cell substrates for the production of both viral vaccines and other biologicals, such as monoclonal antibodies and products prepared using recombinant DNA technology. These requirements specifically exclude DNA vaccines manufactured in microbial cells. Some of the general manufacturing requirements given here (see sections A.2 and A.3) are also applicable to primary cell substrates. Specific requirements for primary cell cultures can be found in the relevant requirements published by WHO (e.g. production of oral poliomyelitis vaccine in primary monkey kidney cells (*25*)).

Whenever practicable, manufacturers are encouraged to use cell substrates that can be generated from master cell banks that have been thoroughly characterized.

Requirements published by WHO are intended to be scientific and advisory in nature. The parts of each section printed in normal type have been written in the form of requirements so that, should a national control authority so desire, they may be adopted as they stand as the basis of national requirements. The parts of each section printed in small type are comments or recommendations for guidance.

Part A. General manufacturing requirements applicable to all types of cell culture production

A.1 **Definitions**

Cell bank: A cell bank is a collection of ampoules containing material of uniform composition stored under defined conditions, each ampoule containing an aliquot of a single pool of cells.

Cell seed: A quantity of well characterized cells of human, animal or other origin stored frozen at -100° C or below in aliquots of uniform composition derived from a single tissue or cell, one or more of which would be used for the production of a master cell bank.

Master cell bank: A quantity of fully characterized cells of human, animal or other origin stored frozen at -100° C or below in aliquots of uniform composition derived from the cell seed. The master cell bank is itself an aliquot of a single pool of cells generally prepared from a selected cell clone under defined conditions, dispensed into multiple containers and stored under defined conditions. The master cell bank is used to derive all working cell banks. The testing performed on a replacement master cell bank (derived from the same cell clone, or from an existing master or working cell bank) is the same as for the initial master cell bank, unless a justified exception is made.

Working cell bank: A quantity of cells of uniform composition derived from the master cell bank at a finite passage level, dispensed in aliquots into individual containers appropriately stored, usually frozen at -100° C or below, one or more of which would be used for production purposes. All containers are treated identically and, once removed from storage, are not returned to the stock.

Production cell cultures: A collection of cell cultures used for biological production that have been prepared together from one or more containers from the working cell bank or, in the case of primary cell cultures, from the tissues of one or more animals.

Adventitious agents: Contaminating microorganisms of the cell culture or line including bacteria, fungi, mycoplasmas and viruses that have been unintentionally introduced.

In vitro culture age: Duration between the thawing of the master cell bank container(s) and the harvest of the production vessel's cell culture as measured by elapsed chronological time in culture, by the population doubling level of the cells, or by the passage level of the cells when subcultivated by a defined procedure for dilution of the culture.

A.2 Good manufacturing practices

The general manufacturing requirements contained in Good Manufacturing Practices for Pharmaceutical (*26*) and Biological (*27*) Products shall apply. Where open manipulations of cells are performed, simultaneous open manipulations of other cell lines shall be avoided to prevent cross-contamination.

Cell cultures shall be prepared by staff who have not, on the same working day, handled animals or infectious microorganisms. The personnel concerned shall be periodically examined medically and found to be healthy.

Particular attention shall be given to the recommendations in Good Manufacturing Practices for Biological Products (*27*) regarding the training and experience of the staff in charge of production and testing and of those assigned to various areas of responsibility in the manufacturing establishment, as well as to the registration of such personnel with the national control authority.

Penicillin or other beta-lactam antibiotics shall not be present in production cell cultures.

> Minimal concentration of other antibiotics may be acceptable. However, the presence of any antibiotic in a biological process or product is discouraged.

A.2.1 *Selection of source materials*

For all types of cells, the donor shall be free of communicable diseases or diseases of uncertain etiology, such as Creutzfeldt-Jakob disease for humans and bovine spongiform encephalopathy (BSE) for cattle.

The national control authority may allow specific exceptions concerning donor health (e.g. myeloma and other tumour cells).

Cells of neurological origin may contain or be capable of amplifying the agent causing spongiform encephalopathies, and shall not be used in the manufacture of medicinal products, apart from cases for which a reasoned exception has been made (*28*).

The national control authority shall approve source(s) of animal-derived raw materials, such as serum and trypsin. These materials shall comply with the guidelines given in the *Report of a WHO Consultation on Medicinal and other Products in relation to Human and Animal Transmissible Spongiform Encephalopathies* (*29*). They shall be subjected to appropriate tests for quality and freedom from contamination by viruses, fungi, bacteria and mycoplasmas to evaluate their acceptability for use in production.

The reduction and elimination from the manufacturing process of raw materials derived from animals and humans is encouraged where feasible.

For some animal-derived raw materials used in the cell culture medium, such as insulin or transferrin, validation of the production process for the elimination of viruses can substitute for virus detection tests.

A.3 Tests applicable to all types of cell cultures

A.3.1 *Tests for viral agents*

Tests shall be undertaken to detect, and where possible identify, any endogenous or exogenous viral agents that may be present in the cells. Special attention shall be given to tests for agents known to be latent in the species from which the cells were derived (e.g. simian virus 40 in rhesus monkeys).

For primary cell cultures, the principles and procedures outlined in Part C, Requirements for Poliomyelitis Vaccine (Oral) (*25*), together with those in section A.4 of Requirements for Measles, Mumps and Rubella Vaccines and Combined Vaccine (Live) (*30*) may be followed. For continuous cell lines and diploid cell substrates see parts B and C below.

A.3.2 *Serum used in cell-culture media*

Serum used for the propagation of cells shall be tested to demonstrate freedom from cultivable bacteria, fungi and mycoplasmas, as specified in Part A, sections 5.2 (*31*) and 5.3 (*32*) of the revised Requirements for Biological Substances No.6 (General Requirements for the Sterility of Biological Substances), and from infectious viruses.

Suitable tests for detecting viruses in bovine serum are given in Appendix 1 of the revised Requirements for Biological Substances No. 7 (Requirements for Poliomyelitis Vaccine, Oral) (*25*). Where appropriate, more sensitive methods may be used.

In some countries, sera are also examined for freedom from certain phages.

In some countries, irradiation is used to inactivate potential contaminant viruses.

The acceptability of the source(s) of serum of bovine origin shall be approved by the national control authority (see A2.1).

Human serum shall not be used. If human albumin is used, it shall meet the revised Requirements for Biological Substances No. 27 (Requirements for the Collection, Processing and Quality Control of Blood, Blood Components and Plasma Derivatives) (*33*), as well as the guidelines contained in *Report of a WHO Consultation on Medicinal and other Products in relation to Human and Animal Transmissible Spongiform Encephalopathies* (*29*).

A.3.3 *Trypsin used for preparing cell cultures*
Trypsin used for preparing cell cultures shall be tested and found free of cultivable bacteria, fungi, mycoplasmas and infectious viruses, especially bovine or porcine parvoviruses, as appropriate. The methods used to ensure this shall be approved by the national control authority.

The source(s) of trypsin of bovine origin shall be approved by the national control authority (see A.2.1).

In some countries, irradiation is used to inactivate potential contaminant viruses.

A.3.4 *Tests for bacteria, fungi and mycoplasmas at the end of production*
A volume of 20 ml of the pooled supernatant fluids from the production cell cultures shall be tested for bacteria, fungi and mycoplasmas as specified in Part A, sections 5.2 (*31*) and 5.3 (*32*) of the Requirements for Biological Substances No.6 (General Requirements for the Sterility of Biological Substances), by a method approved by the national control authority.

A.3.5 *Tests for adventitious viruses at the end of production*
For virus-based products, control cell cultures are necessary when the product interferes with the test systems used to monitor the absence of adventitious agents. These control cell cultures shall be observed at the end of the production period for viral cytopathic effects and tested for haemadsorbing viruses. If multiple harvest pools are prepared at different times, the cultures shall be observed and tested at the time of the collection of each pool.

In some countries, 25% of the control cell cultures are tested for haemadsorbing viruses using guinea-pig red cells. If the red cells have been stored, the duration of storage should not have exceeded 7 days, and the temperature of storage should have been in the range 2-8 °C. In tests for haemadsorbing viruses, calcium and magnesium ions should be absent from the medium.

In some countries, the national control authority also requires that other types of red cells, including cells from humans (blood group IV O), monkeys and chickens (or other avian species) should be used in addition to guinea-pig cells. In all tests, readings should be taken after incubation for 30 minutes at 0-4° C and again after a further incubation for 30 minutes at 20-25° C. For the test with monkey red cells, readings should also be taken after a final incubation for 30 minutes at 34-37° C.

For recombinant DNA proteins, monoclonal antibodies and other cell-based products, the unprocessed bulk harvest or a lysate of cells and their production culture medium shall be tested.

At the time of production of each unprocessed bulk pool, an appropriate volume of the pool shall be inoculated onto monolayer cultures of at least the following cell types:

- Cultures (primary or continuous cell line) of the same species and tissue type as that used for production. This may not be possible for some continuous cell lines (e.g. hybridomas).
- Cultures of a human diploid cell line.
- Cultures of another cell line from a different species.

The unprocessed bulk-pool sample to be tested shall be diluted as little as possible. Material from at least 10^7 cells and spent culture fluids shall be inoculated onto each of the three cell types. The resulting co-cultivated cell cultures shall be observed for evidence of adventitious viruses for at least 2 weeks. If the product is from a continuous cell line known to be capable of supporting the growth of human cytomegalovirus, human diploid cell cultures shall be observed for at least 4 weeks.

Extended cell culture for the purposes of identifying human cytomegalovirus can be replaced by the use of specific probes to detect cytomegalovirus nucleic acid.

At the end of the observation period, aliquots of each of the three co-cultivated cell culture systems shall be tested for haemadsorbing viruses.

Part B. Requirements for continuous-cell-line substrates

B.1 General considerations

Several types of continuous cell line have been employed as substrates in the production of biologicals, including Vero cells in the preparation of live and inactivated viral vaccines and the use of CHO cells in the production of a number of recombinant proteins. The advantage of such cell lines is that they grow relatively rapidly and provide high yields of monolayer or, in some cases, suspension cultures.

Continuous cell lines may have biochemical, biological and genetic characteristics that differ from primary or diploid cells. In particular, they may produce transforming proteins and may contain potentially oncogenic DNA. In some cases, continuous cell lines may cause tumours when inoculated into animals. The manufacturing process for the production of biologicals in continuous-cell-line substrates should take these factors into account in order to ensure the safety of the product. Generally, purification procedures will result in the extensive removal of cellular DNA, other cellular components and potential adventitious agents. Procedures that extensively degrade or denature DNA might be appropriate for some products (e.g. rabies vaccine). When continuous cell lines are being contemplated for use in the development of live viral vaccines, careful consideration must be given to the possible incorporation of oncogenic cellular DNA into the virions.

Production of biologicals from continuous-cell-line substrates should be based on well defined master and working cell banks. The master cell bank is generally derived from a selected cell clone. The working cell bank is derived by expansion of one or more containers of the master cell bank.

Evidence that the cell line is free from cultivable bacteria, mycoplasmas, fungi and infectious viruses, and where appropriate, potentially oncogenic adventitious agents should be provided. Special attention should be given to viruses that commonly contaminate the animal species from which the cell line is derived. Cell seed should preferably be free from all adventitious agents. However, certain cell lines express endogenous viruses, e.g. retroviruses. Tests capable of detecting such agents should be carried out on cells grown under production conditions, and the results should be reported. Specific contaminants identified as endogenous agents in the master and working cell banks should be shown to be inactivated and/or removed by the purification procedure used in production. The validation of the purification procedure used is also considered essential (*34*) (see Appendix).

The data required for the characterization of any continuous cell line to be used for the production of biologicals include: a history of the cell line and a detailed description of the production of the cell banks, including methods and reagents used during culture, *in vitro* culture age, and storage conditions; the results of tests for infectious agents; distinguishing features of the cells, such as biochemical, immunological or cytogenetic patterns which allow them to be clearly distinguished from other cell lines; and the results of tests for tumorigenicity, including data from the scientific literature.

Special consideration should be given to products derived from cells that contain known viral genomes (e.g. Namalva cells). Cells modified by recombinant DNA technology have been increasingly used in the manufacture of novel medicinal products and specific considerations for those products are addressed elsewhere (*35*, *36*).

Continuous cell lines should be characterized so that appropriate controls for the purity and safety of the final product can be included. For example, if a continuous cell line contains an endogenous virus, tests to ensure the absence of any detectable biological activity of that virus could be incorporated as one of the requirements for products derived from that cell line. Alternatively, process validation may replace testing at the end of production for endogenous viruses when a high degree of assurance of consistency of virus clearance can be provided.

There has been considerable discussion internationally on general criteria for the acceptability of products (e.g. hormones, blood components, viral vaccines) prepared from continuous cell lines. A consensus has emerged on the general desirability of achieving a high degree of purification of the product, involving significant removal or destruction of DNA of cell substrate origin. Manufacturers considering the use of continuous cell lines should be aware of the need to develop and evaluate efficient methods for purification as an essential element of any product development programme.

While all continuous cell lines, by definition, have an infinite life span, they may express no tumorigenic properties below a certain passage (or population-doubling) level, but subsequently display increasing evidence of the tumorigenic phenotype with increasing passage. It is therefore important to establish an age limit for *in vitro* cultures beyond which they cannot be used for production. The limit should be based on data derived from production cells expanded under pilot plant-scale or full-scale conditions to the proposed *in vitro* culture age limit or beyond. Generally, the production cells are obtained by expansion of the working cell bank; however, the master cell bank could be used to prepare the production cells, given appropriate justification. Increases in the established *in vitro* culture age limit for production should be supported by data from cells that have been expanded to an *in vitro* culture age that is equal to or greater than the proposed new limit.

The following Requirements concern the characterization and testing of continuous cell lines used for the production of biologicals. They should be read in conjunction with the general manufacturing requirements applicable to all cell cultures contained in part A of these Requirements. Specific requirements for purity as well as other quality control procedures will be incorporated in requirements published by WHO for individual biological products.

B.2 Manufacturing requirements

B.2.1 *Certification of continuous cell lines for use in the production of biologicals*

A continuous cell line used for biologicals production shall be approved by the national control authority and shall be identified by historical records that include information on the origin of the cell line, its method of development and the *in vitro* culture age limit for production.

A continuous cell line used for biologicals production shall also be characterized with respect to genealogy, genetic markers (e.g. histocompatibility leukocyte antigen (HLA), DNA fingerprinting), viability during storage, and growth characteristics at passage levels (or population doublings or time-in-culture, as appropriate) equivalent to, or beyond, those of the master and working cell banks and the cell cultures used for production.

B.2.2 *Cell banks*

The use of continuous cell lines for the manufacture of biological products shall be based on the cell bank system, which shall include a well defined master cell bank and working cell bank.

The cell bank used for the production of biologicals shall be that approved by and registered with the national control authority. The continuous cell line from which the master cell bank has been derived shall be characterized as described in section B1. The working cell bank shall be shown to yield cell cultures capable of producing biologicals that are both safe and efficacious in humans.

In section B.2.3, extensive testing directed at identifying exogenous and endogenous agents that may be present in the cell line is described; special attention is given to agents known to be present in a latent state in the species from which the cells were derived. Such extensive testing need only be performed once, on either the master cell bank or a working cell bank. Once a continuous cell line has been characterized in this respect further testing of working cell banks or production cell cultures is restricted to tests directed at detecting common adventitious agents that could have contaminated the cultures during their preparation.

The tumorigenicity testing described in section B.2.3.7 shall be performed only once on cells of either the master cell bank or a working cell bank propagated to an *in vitro* culture age at or beyond the limit for production. If the cell line has already been documented to be tumorigenic or if the class of cells to which it belongs (e.g., hybridomas) is tumorigenic, the cell line may be presumed to be tumorigenic and tumorigenicity tests need not be undertaken.

Both the master and working cell banks shall be stored at -100° C or below (i.e., in either the liquid or vapour phase of liquid nitrogen). The location, identity and inventory of individual ampoules of cells shall be thoroughly documented.

> It is recommended that the master and working cell banks should each be stored in at least two widely separated areas within the production facility in order to avoid accidental loss of the cell line.

B.2.3 *Identification and characteristics of continuous cell lines*

The characterization of a continuous cell line intended for use in the manufacture of biologicals shall include information on: the history and general characteristics of the cell line; the cell bank system; and quality control testing. These data shall be made available to the national control authority.

B2.3.1 *Identity test*
The cell banks shall be identified by a method approved by the national control authority.

> Methods for identity testing include, but are not limited to, biochemical (e.g., isoenzyme analyses), immunological (e.g., HLA. assays), cytogenetic tests (e.g., for chromosome markers) and tests for genetic markers (DNA fingerprinting).

B.2.3.2 *Sterility tests*
A volume of 20 ml of supernatant fluids from cell cultures derived from at least one ampoule of the master and working cell banks shall be tested for bacteria, fungi and mycoplasmas. Tests shall be performed as specified in Part A, sections 5.2 (*31*) and 5.3 (*32*) of the revised Requirements for Biological Substances No.6 (General Requirements for the Sterility of Biological Substances, by a method approved by the national control authority.

B.2.3.3 *Tests for viral agents using cell cultures*
Live or disrupted cells and spent culture fluids of the master or working cell bank shall be inoculated onto monolayer cultures or cocultivated with monolayer cultures, as appropriate, of the following cell types:

- Cultures (primary or continuous cell line) of the same species and tissue type as the continuous cell line. This may not be possible for some continuous cell lines, e.g. hybridomas.
- Cultures of a human diploid cell line.
- Cultures of another cell line from a different species.

The sample to be tested shall be diluted as little as possible. Material from at least 10^7 cells and spent culture fluids shall be inoculated onto each of the three cell types. The resulting cultures shall be observed for at least 2 weeks for evidence of adventitious viruses. If the continuous cell line being tested is known to be capable of supporting the growth of human cytomegalovirus, human diploid cell cultures shall be observed for at least 4 weeks.

> Extended cell culture for the purposes of identifying human cytomegalovirus can be replaced by the use of specific probes to detect cytomegalovirus nucleic acid.

At the end of the observation period, aliquots of each of the three cell culture systems shall be tested for haemadsorbing viruses.

B.2.3.4 *Tests for viral agents using animals and eggs*
The cells of the master and working cell banks are suitable for production if none of the animals or eggs shows evidence of the presence of any viral agent attributable to the cell banks.

Tests in animals. Tests in animals for pathogenic viruses shall include the inoculation by the intramuscular route of each of the following groups of animals with cells from the master or working cell banks, propagated to or beyond the maximum *in vitro* culture age (or population doubling, as appropriate) used for production, where at least 10^7 viable cells are divided equally among the animals in each group:

- two litters of suckling mice, comprising a total of at least ten animals, less than 24 h old; and
- ten adult mice weighing 15-20 g.

In some circumstances, tests in five guinea-pigs weighing 350-450 g and five rabbits weighing 1.5-2.5 kg may be considered.

The test in rabbits tor the presence of B virus in cell lines of simian origin may be replaced by a test in rabbits kidney-cell cultures.

The animals shall be observed for at least 4 weeks. Any animals that are sick or show any abnormality shall be investigated to establish the cause. The test is not valid if more than 20% of the animals in the test group become sick for non-specific reasons and do not survive the observation period.

In some countries, the suckling and adult mice are also inoculated by the intracerebral route.

If the cell line is of rodent origin, at least 10^6 viable cells shall be injected intracerebrally into each of ten susceptible adult mice to test for the presence of lymphocytic choriomeningitis virus.

Tests in eggs. At least 10^6 viable cells from the master or working cell banks, propagated to or beyond the maximum *in vitro* culture age (or population doubling, as appropriate) shall be injected into the allantoic cavity of each of ten embryonated chicken eggs, and the yolk sac of each of another ten embryonated chicken eggs. The eggs shall be examined after not less than 5 days of incubation. The allantoic fluids of the eggs shall be tested with red cells from guinea-pig and chickens (or other avian species) for the presence of haemagglutinins. The test is not valid if more than 20% of the embryonated chicken eggs in the test group are discarded for non-specific reasons.

Usually, the eggs used for the yolk sac test should be 5-6 days old. The eggs used for the allantoic cavity test should be 9-11 days old.

Alternative ages for the embryonated chicken eggs and alternative incubation periods are acceptable if they have been determined to be capable of detecting the presence of routine adventitious agents in the test samples.

B.2.3.5 *Tests for retroviruses and other endogenous viruses or viral nucleic acid*
Test samples from the master or working cell banks, propagated to or beyond the maximum *in vitro* culture age (or population doubling, as appropriate) shall be examined for the presence of retroviruses using the following techniques:

- infectivity assays (if the infectivity assay is positive, tests for reverse transcriptase are not necessary);
- transmission electron microscopy (TEM); and
- reverse transcriptase (RTase) assays (performed in the presence of magnesium and manganese) on pellets obtained from fluids by high speed centrifugation (e.g. 125000g for 1 h) at 4°C.

Recently developed highly sensitive RTase assays may be considered, but the results need to be interpreted with caution because RTase activity is not unique to retroviruses and may derive from other sources, such as retrovirus-like elements which do not encode a complete genome or cellular DNA polymerase.

It is often possible to increase the sensitivity of cell-culture infectivity assays by first inoculating the test material onto human cell lines that can support retroviral growth in order to amplify any retrovirus contaminant that may be present at low concentrations. For non-murine retroviruses, test cell lines should be selected for their capacity to support the growth of a broad range of retroviruses, including viruses of human and non-human primate origin (*37*, *38*).

For murine retroviruses, amplification of low-level contaminants may be achieved by co-cultivation of cells with a highly susceptible cell line, e.g. *Mus dunni* cells (*39*). The latter are susceptible to infection by all tested murine leukaemia viruses except Moloney murine leukaemia virus. For that reason, another susceptible cell, for example SC-1 (*40*), should also be used. Fluid from the resulting co-cultures should be further passaged on *Mus dunni* or other susceptible cells and subsequently assayed for murine leukaemia virus.

A variety of other assays may be useful, depending on the circumstances. Some examples of such assays include viable cell immunofluorescence (IFA) on *Mus dunni* cells co-cultivated with the test cells using a broadly reactive monoclonal antibody (e.g., HY95) for the detection of ecotropic, xenotropic, mink-cell focus-forming and amphotropic viruses; feline S + L assays using PG4 cells (*41*) for detection of amphotropic viruses; mink S + L assays for detection of xenotropic viruses (*12*) and mouse S + L assays using D56 (*42*) cells for detection of ecotropic viruses.

Murine and other rodent cell lines or hybrid cell lines containing a rodent component should be assumed to be inherently capable of producing infectious retroviruses. For murine cell lines used for monoclonal antibody production, the extent of testing for specific retroviruses may be reduced. However, the manufacturing process should be evaluated for removal and/or inactivation of retroviruses. For murine-human hybrid cell lines, additional concerns arise. Any proposed testing should be discussed with the national control authority on a case-by-case basis.

Probe hybridization/polymerase-chain-reaction amplification and virus-specific monoclonal antibody detection may provide additional information on the presence or absence of specific contaminants.

B.2.3.6 *Tests tor selected viruses*
The following tests shall be undertaken on a selected basis on samples from the working cell bank propagated to or beyond the maximum *in vitro* culture age (or population doubling, as appropriate).

Murine cell lines shall be tested for species-specific viruses using mouse, rat and hamster antibody production tests. *In vivo* testing for lymphocytic choriomeningitis virus, including a challenge for non-lethal strains, is required for such cell lines as specified in B.2.3.4.

Human cell lines shall be screened for human viral pathogens such as Epstein-Barr virus, cytomegalovirus, human retroviruses, and hepatitis B and C viruses with appropriate *in vitro* techniques. Selection of the viruses to be screened for shall take into account the tissue source and medical history of the person from whom the cell line was derived. Tests for retroviruses are specified in section B.2.3.5.

The use of other cell cultures also may be appropriate for the characterization of cell banks, depending on the cell type and source of the cell line being characterized (*17*). Under certain circumstances, specific testing tor the presence of other transforming viruses, such as papillomavirus, adenovirus and herpesvirus 6 and 7, may also be indicated.

B.2.3.7 *Tests for tumorigenicity*
If the continuous cell line has already been demonstrated to be tumorigenic (e.g., BHK21, CHO, C127), or if the class of cells to which it belongs, for example hybridoma, is tumorigenic, it is not necessary to require additional tumorigenicity tests. A new cell line shall be presumed to be tumorigenic unless data demonstrate that it is not. If a manufacturer proposes to characterize the cell line as non-tumorigenic, the following tests shall be undertaken.

Tests in vivo. Cells from the master or working cell bank propagated to or beyond the *in vitro* culture age limit for production shall be examined for tumorigenicity in a test approved by the national control authority. The test shall involve a comparison between the continuous cell line and a suitable positive reference preparation (e.g., HeLa, Hep 2 or FL cells).

A negative control is not essential but desirable. For that purpose non-tumorigenic diploid cell lines such as WI-38 or MRC-5 may be used.

Animal systems that have been shown to be suitable for this test include:

(a) athymic mice (*Nu/Nu* genotype);
(b) newborn mice, rats or hamsters that have been treated with antithymocyte serum or globulin; and
(c) thymectomized and irradiated mice that have been reconstituted (T-, B+) with bone marrow from healthy mice.

Whichever animal system is selected, the cell line and the reference cells are injected into separate groups of ten animals each. In both cases, the inoculum for each animal is 10^7 cells suspended in a volume of 0.2ml, and the injection is by either the intramuscular or the subcutaneous route. In the case of newborn animals (b), the animals are treated with 0.1ml of antithymocyte serum or globulin on days 0, 2, 7 and 14 after birth. A potent serum or globulin is one that suppresses the immune mechanisms of the growing animals to the extent that the subsequent inoculum of 10^7 positive reference cells regularly produces tumours and metastases.

At the end of the observation period all animals, including the reference group(s), shall be killed and examined for gross and microscopic evidence of the proliferation of inoculated cells at the site of injection and in other organs (e.g. lymph nodes, lungs, kidneys and liver).

In all test systems, the animals shall be observed and palpated at regular intervals for the formation of nodules at the sites of injection. Any nodules formed should be measured in two perpendicular dimensions, the measurements being recorded regularly to determine whether there is progressive growth of the nodule. Animals showing nodules which begin to regress during the period of observation shall be killed before the nodules are no longer palpable, and processed for histological examination. Animals with progressively growing nodules shall be observed for 1-2 weeks. Among those without nodule formation, half shall be observed for 3 weeks and half for 12 weeks before they are killed and processed for histological examination. A necropsy shall be performed on each animal and shall include examination for gross evidence of tumour formation at the site of inoculation and in other organs such as lymph nodes, lungs, brain, spleen, kidneys and liver. All tumour-like lesions and the site of inoculation shall be examined histologically. In addition, since some cell lines may give rise to metastases without evidence of local tumour growth, any detectable regional lymph nodes and the lungs of all animals shall be examined histologically.

For the test to be considered valid, progressively growing tumours must be produced in at least nine of ten animals injected with the positive reference cells.

In vitro tests may be considered sufficient by some national control authorities.

Two *in vitro* tests that have been found to provide useful additional information on tumorigenicity are: colony formation in soft agar gels, and production of invasive cell growth following inoculation onto organ cultures. They may be used to characterize more fully the cell lines that show no evidence of tumorigenicity in animal tests (see above), or when the results are equivocal.

As the cells used in the production of biologicals may contain activated oncogenes, assays of cell transformation with DNA derived from a continuous cell line at the limit for *in vitro* culture age for production should be considered in order to determine whether or not activated oncogenes can be detected. The 3T3 assay system has been found useful for *ras* assays. Additional tests may also be considered as new techniques are developed for the detection of a broader range of oncogenes.

B.2.3.8 *Tests on cells carrying a recombinant-DNA expression system*
Data shall be obtained demonstrating that a continuous cell line can be used for its intended purpose. If a continuous cell line contains an expression construct to produce a recombinant DNA-derived protein, data shall be obtained to demonstrate the consistent quality and quantity of the protein it produces throughout the proposed *in vitro* culture age range for production (*14*, *15*). Studies shall be performed to determine whether manipulation of the cell line in order to produce a product by transfection changes its biological characteristics significantly, for instance conversion to the tumorigenic phenotype. Any such change must be taken into account in product development and in assessing approaches taken to assure an acceptable product.

The International Conference on Harmonisation has issued additional useful information (*43*).

B.2.4 *Production cell cultures*
Characterization of the product and routine monitoring for adventitious agents during the production process are part of the quality control of biological products.

The choice of method for quality control of the production cell substrate depends on the nature of the propagation system used. Cell substrates are propagated as monolayer cultures, in suspension cultures or in bioreactors, and can be maintained on a short-term, a long-term or even on a potentially indefinite basis. The product is obtained either from a single harvest of cell culture fluid or from multiple harvests. In some cases, quality control testing may need to be performed on each harvest before pooling into a bulk lot. The management of cell substrates for the purposes of quality control testing should be designed to optimize sensitivity of the testing.

B.2.4.1 *Serum used in cell-culture media*
Serum used in cell-culture media shall be tested as specified in section A.3.2.

B.2.4.2 *Trypsin used for preparing cell cultures*
Trypsin used for preparing cell cultures shall be tested as specified in section A.3.3.

B.2.4.3 *Identity test*
For viral vaccines, an identity test shall be performed on the control cell culture as described in section B.2.3.1. For recombinant DNA proteins and monoclonal antibodies, the presence of the protein at consistent levels in the harvest is an adequate confirmation of identity and purity.

B.2.4.4 *Tests for bacteria, fungi and mycoplasmas at the end of production*
Tests for bacteria, fungi and mycoplasmas shall be conducted on the production culture supernatant or lysate as specified in section A.3.4.

B.2.4.5 *Tests for adventitious viruses at the end of production*
Tests for adventitious viruses shall be conducted on the production culture supernatant or lysate as specified in section A.3.5.

Part C. Requirements for diploid cell substrates

C.1 General considerations

Two human diploid cell lines, WI-38 and MRC-5, derived from embryo lung tissue, have been in widespread use for many years for the production of live virus vaccines, including oral poliomyelitis, measles, mumps, rubella and varicella vaccines, and inactivated vaccines, for example, rabies and hepatitis A vaccines. In addition, a rhesus diploid cell line, FRhL-2, has been in limited use for rabies vaccine production. These substrates have been found to be safe and to produce vaccines that stimulate effective immunity without untoward reactions attributable to the cell substrate.

The following requirements concern the characterization and testing of diploid cell lines used for the production of biologicals. They should be read in conjunction with the general manufacturing requirements applicable to all cell cultures contained in Part A of these Requirements.

C.2 Manufacturing requirements

C.2.1 *Certification of diploid cell lines for use in the production of biologicals*

A diploid cell line used for biologicals production shall be approved by the national control authority and shall be identified by historical records that include information on the origin of the cell line, its method of development and the range of passage levels at which it can be used in biologicals production.

A new diploid cell line (e.g. other than WI-38, MRC-5 and FRhL-2) used for biologicals production shall be characterized with respect to genealogy, genetic markers (e.g. HLA, DNA fingerprinting), or other markers of identity acceptable to the national control authority, as well as for viability during storage. In addition, data must be obtained to establish the cell line's diploid character and growth characteristics at *in vitro* culture ages equivalent to, or beyond, those of the master and working cell banks, and of the cell cultures used for production.

Accumulated experience suggests that WI-38 and MRC-5 can be used for production until 10 generations before senescence.

C.2.2 *Cell banks*

C.2.2.1 *Master cell bank and working cell bank*

Tests shall be performed on the master and working cell banks as described in section C.2.3, where appropriate and approved by the national control authority. In addition, for a new diploid cell line (e.g., other than WI-38, MRC-5, and FRhL-2) the cells of the working cell bank shall be shown to be diploid and stable with respect to karyology by the tests outlined in section C.2.3.5.

C.2.3 *Identification and characteristics of diploid cell lines*

The characterization of a diploid cell intended for use in the manufacture of biologicals shall include information on: the history and general characteristics of the cell line; the cell bank system; and quality control testing. These data shall be made available to the national control authority.

C.2.3.1 *Identity tests*

An identity test shall be performed on the master cell bank by a method approved by the national control authority.

Methods for identity testing include, but are not limited to, biochemical tests (e.g., isoenzyme analyses), immunological tests (e.g. HLA assays), cytogenetic tests (e.g., for chromosomal markers), and tests for genetic markers (DNA fingerprinting).

Tests to ensure that the master cell bank is not contaminated with a continuous cell line shall be performed.

Tests of identity such as DNA fingerprinting of appropriate sensitivity, karyology at different levels of passage or studies of lifespan in culture may be used for this purpose if approved by the national control authority.

C.2.3.2 *Sterility tests*

Tests for bacteria, fungi and mycoplasmas shall be conducted in cell cultures as specified in section B.2.3.2.

C.2.3.3 *Tests for viral agents using cell cultures*

Tests for viral agents shall be conducted in cell cultures as specified in section B.2.3.3.

C.2.3.4 *Tests for viral agents using animals and eggs*

Tests for viral agents shall be conducted in animals and eggs as specified in section B.2.3.4.

C.2.3.5 *Chromosomal characterization of a diploid cell line*
The usefulness of chromosomal characterization depends on the nature of the product and the manufacturing process. In general, products that might contain live cells or which have little "down-stream" purification will require chromosomal characterization of the cell line. Such products manufactured in cells identified to be WI-38, MRC-S or FRhL-2 cells do not require re-characterization of the cell substrate by karyology, unless the cells have been genetically modified.

The utility of chromosomal monitoring of the cell substrate for unpurified products manufactured in other cell lines shall be evaluated on a case-by-case basis. However, products that contain no cells and are highly purified will not require this test.

For the determination of the general character of a new diploid cell line (i.e., other than WI-38, MRC-5 and FRhL-2), samples from the master cell bank shall be examined at approximately four equally spaced intervals over the life span of the cell line during serial cultivation through to senescence. Each sample shall consist of a minimum of 200 cells in metaphase and shall be examined for exact counts of chromosomes and for frequency of hyperdiploidy, hypoploidy, polyploidy, breaks and structural abnormalities. The acceptability of any new diploid cell line shall be determined by the national control authority.

> It is recommended that photographic reconstruction should be employed to prepare chromosome-banded karyotypes of an additional ten metaphase cells.

Stained slide preparations of the chromosomal characterization of the diploid cell line, or photographs of these, shall be maintained permanently as part of the cell line record.

C.2.3.7 *Tests for tumorigenicity*
The tumorigenic potential of a new diploid cell line (i.e., other than WI-38, MRC-5 and FRhL-2) shall be tested as specified in section B.2.3.7 as part of the characterization of the cell line, but is not required on a routine basis.

> If satisfactory data from at least two independent laboratories are available, further tumorigenicity testing may not be required. The adequacy of tumorigenicity testing of a new diploid cell line should be discussed with the national control authority. Positive results should be discussed with the authority, taking into consideration the purity of the product, including residual cellular DNA.

C.2.3.8 *Tests on cells carrying a recombinant-DNA expression construct*
Data shall be obtained demonstrating that a diploid cell line can be used for its intended purpose. If a cell line contains an expression construct to produce a recombinant-DNA-derived protein, data shall be obtained to demonstrate the consistent quality and quantity of the protein produced throughout the proposed *in vitro* culture age range for production (*33*, *34*).

The International Conference on Harmonisation has issued additional useful information (*43*).

C.2.4 Production cell cultures

C.2.4.1 *Serum used in cell-culture media*

Serum used in cell-culture media shall be tested as specified in section A.3.2.

C.2.4.2 *Trypsin used for preparing cell cultures*

Trypsin used for preparing cell cultures shall be tested as specified in section A.3.3.

C.2.4.3 *Identity test*

An identity test shall be performed on the control cell culture as specified in section A.2.3.1.

C.2.4.4 *Tests for bacteria, fungi and mycoplasmas at the end of production*

Tests for bacteria, fungi and mycoplasmas shall be conducted on the production culture supernatant or lysate as specified in section A.3.4.

C.2.4.5 *Tests for adventitious viruses at the end of production*

Tests shall be conducted on the product at the end of production but before further processing as specified in section A.3.5. If the presence of the product interferes, tests shall be performed on the control cell culture as specified in section A.3.5.

Authors

The first draft of these Requirements was prepared by Dr V. Grachev, Scientist, Biologicals, Dr D. Magrath, Chief (1987-1994), Biologicals, and Dr E. Griffiths, Chief (from 1994), Biologicals, World Health Organization, Geneva, Switzerland.

A revised draft was formulated by Dr V. Grachev, Deputy Director, Institute of Poliomyelitis and Viral Encephalitides, Russian Academy of Medical Sciences, Moscow, Russian Federation, Dr E. Griffiths, Chief, Biologicals, WHO, Geneva, Switzerland and Dr JC Petricciani, Vice-President, Genetics Institute, Cambridge, MA, USA.

After extensive consultation, including discussion of the main scientific issues at a WHO/IABS meeting held at the Mérieux Foundation, Annecy, France, from 29 September to 1 October 1996, further amendments to the text were proposed by the following group of experts:

Dr Y.Y. Chiu, Director, Division of New Drug Chemistry, Food and Drug Administration, Rockville, MD, USA
Dr R. Dobbelaer, Acting Chief, Biological Standardization, Institute for Hygiene and Epidemiology, Brussels, Belgium
Dr V. Grachev, Institute of Poliomyelitis and Viral Encephalitides, Russian Academy of Medical Sciences, Moscow, Russian Federation
Dr E. Griffiths, Chief, Biologicals, World Health Organization, Geneva, Switzerland
Dr I. Gust, CSL Limited, Parkville, Victoria, Australia
Dr M.C. Hardegree, Director, Office of Vaccine Research and Review, Center for Biologics Evaluation and Research, Food and Drug Administration, Bethesda, MD, USA
Dr T. Hayakawa, National Institute of Health Science, Tokyo, Japan
Professor F. Horaud, Pasteur Institute, Paris, France
Dr A.S. Lubiniecki, SmithKline Beecham, King of Prussia, PA, USA
Dr P. Minor, Head, Division of Virology, National Institute for Biological Standards and Control, Potters Bar, England
Dr B. Montagnon, Pasteur Mérieux Sera and Vaccines, Marcy l'Etoile, France
Dr J. Peetermans, Smith Kline Beecham Biologicals, Rixensart, Belgium
Dr J. Petricciani, Vice-President, Genetics Institute, Cambridge, MA, USA
Dr A. Ridgeway, Head, Biotechnology Section, Health and Welfare Canada, Ottawa, Ontario, Canada
Dr J. Robertson, National Institute for Biological Standards and Control, Potters Bar, England
Dr G. Schild, Director, National Institute for Biological Standards and Control, Potters Bar, England
Dr K.B. Seamon, Immunex Corporation, Seattle, WA, USA

Acknowledgements

Acknowledgements are due to the following experts for their comments and advice: Dr S.C. Arya, Centre for Logistical Research and Innovation, New Delhi, India; Dr T. Bektimirov, Deputy Director, Tarasevic State Institute for the Standardization and Control of Medical Biological Preparations, Moscow, Russian Federation; Dr M. Duchêne, Director. Quality Control, Smith Kline Beecham Biologicals, Rixensart, Belgium; Dr B.D. Garfinkle, Vice-President, Vaccine Quality Operations. Merck & Co., West Point. PA. USA; Dr R.K. Gupta, Assistant Director, Massachusetts Public Health Biologics laboratory, Boston, MA, USA; Dr K. Healy, Head, Quality Assurance Department, CSL Limited, Parkville, Victoria, Australia; Dr A. Homma, Regional Adviser in Biologics. WHO Regional Office for the Americas, Washington, DC, USA; Dr J.G. Kreeftenberg, Head, Quality and Regulatory Affairs, National Institute of Public Health and Environmental Protection, Bilthoven, Netherlands; Dr L. Lavese, Head of International Affairs Department, Farmindustria, Rome, Italy; Dr R. Netter. Rue Vaugirard, Paris, France; Dr A.S. Outschoorn, Scientist Emeritus, United States Pharmacopeia, Rockville, MD, USA; Dr E. Walker, Head, Molecular Biology Section, Therapeutic Goods Administration, Woden, Australia.

References

1. **Van Wezel AL.** Microcarrier technology - present status and prospects. *Developments in biological standardization*, 1984, **55**:3.
2. **Jacobs JP et al.** Guidelines for the acceptability, management and testing of serially propagated human diploid cells for the production of live virus vaccines for use in man. *Journal of biological standardization*, 1981, **9**:331-342.
3. **Hayflick L, Plotkin S, Stevenson R.** History of the acceptance of human diploid cell strains as substrates for human virus vaccine production. *Developments in biological standardization*, 1987, **68**:9-17.
4. **Grachev, V.** World Health Organization attitude concerning the use of continuous cell lines as substrate for production of human virus vaccines. In: Mizrahi A, ed. *Advances in biotechnological processes. Vol. 14. Viral vaccines.* Wiley-Liss, 1990:37-67.
5. *Acceptability of cell substrates for production of biologicals. Report of a WHO Study Group.* Geneva, World Health Organization, 1987 (WHO Technical Report Series, No. 747).
6. Requirements for continuous cell lines used for biologicals production. In: *WHO Expert Committee on Biological Standardization. Thirty-sixth Report.* Geneva, World Health Organization, 1987, Annex 3 (WHO Technical Report Series. No. 745).

7. WHO bank of Vero cells for the production of biologicals. In: *WHO Expert Committee on Biological Standardization. Fortieth Report.* Geneva, World Health Organization, 1990, p. 11 (WHO Technical Report Series, No. 800).
8. **Temin HM.** Overview of biological effects of addition of DNA molecules to cells. *Journal of medical virology*, 1990, **31**: 13-17.
9. **Kurth R.** Risk potential of the chromosomal insertion of foreign DNA. *Annals of the New York Academy of Sciences*, 1995, **772**:140-150.
10. **Petricciani JC, Horaud FN.** DNA, dragons, and sanity. *Biologicals*, 1995, **23**:233-238.
11. **Nichols WW et al.** Potential DNA vaccine integration into host cell genome. *Annals of the New York Academy of Science*, 1995, **772**:30-38.
12. **Coffin JM.** Molecular mechanisms of nucleic acid integration. *Journal of medical virology*, 1990, **31**:43-49.
13. **Petricciani JC, Regan PJ.** Risk of neoplastic transformation from cellular DNA: calculations using the oncogene model. *Developments in biological standardization*, 1986, **68**:43-49.
14. **Strain AJ.** The uptake and fate of DNA transfected into mammalian cells *in vitro*. *Developments in biological standardization*, 1986, **68**:27-32.
15. **Doehmer J.** Residual cellular DNA as a potential transforming factor. *Developments in biological standardization*, 1986, **68**:33-41.
16. **Wierenga DE, Cogan J, Petriccaini JC.** Administration of tumor cell chromatin to immunosuppressed and non-immunosuppressed primates. *Biologicals*, 1995, **23**:221-224.
17. **Greenwald P et al.** Morbidity and mortality among recipients of blood from pre-leukemic and pre-lymphomatous donors. *Cancer*, 1976, **38**: 324-328.
18. **Fleckenstein B, Daniel MD, Hunt RD.** Tumor inductions with DNA of oncogenic primate herpesviruses. *Nature*, 1978, **274**: 57-59.
19. **Duxbury M et al.** DNA in plasma of human blood for transfusion. *Biologicals*, 1995, **23**:229.
20. **Nawroz H et al.** Microsatellite alterations in serum DNA of head and neck cancer patients. *Nature medicine*, 1996, **2**:1035-1037.
21. **Perrin P, Morgeaux S.** Inactivation of DNA by beta-propiolactone. *Biologicals*, 1995, **23**:207.
22. Requirements for Poliomyelitis Vaccine (Inactivated). In: *Requirements for Biological Substances: 1. General Requirements for Manufacturing Establishments and Control Laboratories; 2. Requirements for Poliomyelitis Vaccine (Inactivated). Report of a Study Group.* Geneva, World Health Organization, 1959 (WHO Technical Report Series, No. 178).
23. Requirements for Poliomyelitis Vaccine (Inactivated). In: *Requirements for Biological Substances. Manufacturing and Control Laboratories. Report of a*

WHO Expert Group. Geneva, World Health Organization, 1966 (WHO Technical Report Series, No. 323).

24. **Hilleman M.** *Cells, vaccines and pursuit of precedent.* Bethesda, MD, National Cancer Institute, 1968 (National Cancer Institute Monograph, No. 29).
25. Requirements for Poliomyelitis Vaccine (Oral) (Requirements for Biological Substances No.7, revised 1989). In: *WHO Expert Committee on Biological Standardization. Fortieth Report.* Geneva, World Health Organization, 1990, Annex 1 (WHO Technical Report Series, No. 800).
26. Good manufacturing practices for pharmaceutical products. In: *WHO Expert Committee on Specifications for Pharmaceutical Preparations. Thirty-second Report.* Geneva, World Health Organization, 1992, Annex 1 (WHO Technical Report Series, No. 823).
27. Good manufacturing practices for biological products. In: *WHO Expert Committee on Biological Standardization. Forty-second Report.* Geneva, World Health Organization, 1992. Annex 1 (WHO Technical Report Series, No. 822).
28. Questions de santé publique liées aux encéphalopathies spongiformes chez l'animal et chez l'homme: Mémorandum d'une reunion de l'OMS. *Bulletin of the World Health Organization*, 1992,70:573-582.
29. *Report of a WHO Consultation on Medicinal and Other Products in Relation to Human and Animal Transmissible Spongiform Encephalopathies.* Geneva, World Health Organization, 1997 (unpublished document WHO/BLG/97.2; available on request from Biologicals, World Health Organization, 1211 Geneva 27, Switzerland).
30. Requirements for Measles, Mumps and Rubella Vaccines and Combined Vaccine (Live) (Requirements for Biological Substances No. 47, 1992). In: *WHO Expert Committee on Bioloqicel Standardization. Forty-fourth Report.* Geneva, World Health Organization, 1994, Annex 3 (WHO Technical Report Series, No. 840).
31. General Requirements for the sterility of Biological Substances (Requirements for Biological Substances No.6, revised 1973). In: *WHO Expert Committee on Biological Standardization. Twenty-fifth Report.* Geneva, World Health Organization, 1973, Annex 4 (Technical Report Series, No. 530).
32. General Requirements for the Sterility of Biological Substances (Requirements for Biological Substances No.6, revised 1973, amendment 1995). In: *WHO Export Committee on Biological Standardization. Forty-sixth Report.* Geneva. World Health Orpanization, 1998, Annex 3 (WHO Technical Report Series, No. 872).
33. Requirements for the Collection, Processing and Quality Control of Blood, Blood Components and Plasma Derivatives (Requirements for Biological Substances No. 27, revised 1992). In: *WHO Expert Committee en Biological Standardization.*

Forty-third Report. Geneva, World Health Organization, 1994, Annex 2 (WHO Technical Report Series No. 840).

34. *Viral safety evaluation of biotechnology products.* Geneva, International Conference on Harmonisation, 1997 (unpublished document available on request from International Conference on Harmonisation Secretariat, International Federation of Pharmaceutical Manufacturers Associations, 30 rue de St-Jean, 1211 Geneva 18, Switzerland).
35. Guidelines for assuring the quality of pharmaceutical and biological products prepared by recombinant DNA technology. In: *WHO Expert Committee on Biological Standardization. Forty-first Report.* Geneva, World Health Organization, 1991, Annex 3 (WHO Technical Report Series, No. 814).
36. Requirements for Hepatitis B Vaccines made by Recombinant DNA Techniques. In: *WHO Expert Committee on Biological Standardization. Thirty-ninth Report.* Geneva, World Health Organization, 1989, Annex 2 (WHO Technical Report Series, No. 786).
37. **Peebles PT.** An *in vitro* focus-induction assay tor xenotropic murine leukemia virus, feline leukemia virus C, and the feline-primate viruses RD-114/CC/M-7. *Virology*, 1975, **67**:288-291.
38. **Sammerfelt MA, Weiss RA.** Receptor interference groups of 20 retroviruses plating on human cells. **Virology**, 1990, **176**:58-69.
39. **Lander MR, Chattopadhyay SK.** A *Mus dunni* cell line that lacks sequences closely related to endogenous murine leukemia viruses and can be infected by ecotropic, amphotropic, xenotropic, and mink cell focus-forming viruses. *Journal of virology*, 1984, **52**:695-698.
40. **Hartley RH, Rowe WP.** Clonal cell lines from a feral mouse embryo which lack host-range restrictions for murine leukemia viruses. *Virology*, 1975, **65**:128-134.
41. **Bassin RH et al.** Normal DBA/2 mouse cells synthesize a glycoprotein which interferes with MCF virus infection. *Virology*, 1982, **123**:139-151.
42. **Bassin RH, Tuttle N, Fischinger PJ.** Rapid cell culture assay technic for murine leukeamia viruses. *Nature*, 1971, **229**:564-566.
43. *Analysis of the expression construct in cells used for production of a DNA-derived protein product.* Geneva, International Conference on Harmonisation, 1995 (unpublished document available on request from International Conference on Harmonisation Secretariat, International Federation of Pharmaceutical Manufacturers Associations, 30 rue de St-Jean, 1211 Geneva 18, Switzerland).

WHO Expert Committee on Biological Standardization

Fifty-fourth Report

World Health Organization
Technical Report Series (TRS)
927

World Health Organization, Geneva **2005**

WHO Expert Committee on Biological Standardization

Geneva, 17-21 November 2003

Members

Dr W.G. van Aken, Amstelveen, the Netherlands (*Chairman*)
Dr D.H. Calam, Pewsey, Wiltsllire, England (*Rapporteur*)
Dr M. de los Angeles Cortes Castillo, Director, Quality Control, National Institute of Hygiene, Mexico City, Mexico
Dr R. Dobbe!aer, Head, Biological Standardization, Scientific Institute of Public Health, Brussels, Belgium (*Vice-Chairman*)
Dr F. Fuchs, Director, Lyon Site, French Agency for Safety of Health Products, Lyon, France
Dr B. Kaligis, International Relations Division Head, Bio Farma, Bandung, Indonesia
Dr N.V. Medunitsin, Director, Tarasevic State Institute for the Standardization and Control of Medical Biological Preparations, Moscow, Russian Federation
Dr F. Ofosu, Department of Pathology and Molecular Medicine, McMaster University, Hamilton, Ontario, Canada
Dr F. Reigel, Head, Biological Medicines and Laboratories, Swissmedic, Agency for Therapeutic Products, Berne, Switzerland

Representatives of other organizations

Developing Country Vaccine Manufacturer's Network
Dr S. Jadhav, Executive Director (Quality Assurance), Serum Institute of India Ltd., Maharashtra, Pune, India
European Diagnostic Manufacturers Association
Dr J. Diment, Scientific and Technical Manager, North Europe, Asia Pacific and Export, Ortho-Clinical Diagnostics, Amersham, England
Council of Europe, European Directorate for the Quality of Medicines
Mr J.-M. Spieser, Head, Division IV Biological Standardisation/Network of Official Medicines Control Laboratories (OMCLs), Strasbourg, France
International Association of Biologicals
Dr A. Eshkol, Senior Scientific Adviser, Aeres Serono International SA, Geneva, Switzerland
International Federation of Clinical Chemistry and Laboratory Medicine
Professor J. Thijssen, c/o Department of Endocrinology, University Hospital Utrecht, Utrecht, the Netherlands
International Federation of Pharmaceutical Manufacturers Associations
Dr M. Duchêne, Director, Quality Control, GlaxoSmithKline Biologicals, Rixensart, Belgium
Dr A. Sabouraud, Director, Quality Control of Development Products, Aventis Pasteur SA, Marcy l'Etoile, France

International Society on Thrombosis and Haemostasis
Professor I.R. Peake, Division of Genomic Medicine, Royal Hallamshire Hospital, Sheffield, England
Plasma Protein Therapeutics Association
Dr E. Hutt, Brussels, Belgium
United States Pharmacopeia
Dr T. Morris, Department of Information and Standards, Development, Rockville, MD, USA

Secretariat
Dr T. Barrowcliffe, Head, Division of Haematology, National Institute for Biological Standards and Control, Potters Bar, Herts., England (*Temporary Adviser*)
Dr A. Bristow, Head, Division of Endocrinology, National Institute for Biological Standards and Control, Potters Bar, Herts., England (*Temporary Adviser*)
Dr M. Corbel, Head, Division of Bacteriology, Division of Bacteriology, National Institute for Biological Standards and Control, Potters Bar, Herts., England (*Temporary Adviser*)
Dr R. Decker, Hepatitis and AIDS Research, Deerfield, Illinois, USA (*Temporary Adviser*)
Dr W. Egan, Acting Director, Office of Vaccines, Center for Biologics, Evaluation and Research, Food and Drug Administration, Rockville, MD, USA (*Temporary Adviser*)
Dr J. Epstein, Director, Office of Blood Research and Review, Center for Biologics Evaluation and Research, Food and Drug Administration, Rockville, MD, USA (*Temporary Adviser*)
Dr I. Feavers, Division of Bacteriology, National Institute for Biological Standards and Control, Potters Bar, Herts., England (*Temporary Adviser*)
Dr E. Griffiths, Associate Director General, Biologics and Genetic Therapies, Ottawa, Ontario, Canada (*Temporary Adviser*)
Dr M.F. Gruber, Scientific Reviewer, Division of Vaccines and Related Products Application, Center for Biologics Evaluation and Research, Food and Drug Administration, Rockville, MD, USA (*Temporary Adviser*)
Dr S. Inglis, Director, National Institute for Biological Standards and Control, Potters Bar, Herts., England (*Temporary Adviser*)
Dr L. Jodar, Associate Director for Programmes, Institutional Development and Partnerships, International Vaccine Institute, Kwanak-Ku, Seoul, Republic of Korea (*Temporary Adviser*)
Dr N. Lelie, Viral Quality Control Unit, Sanquin-CLB, Alkmaar, the Netherlands (*Temporary Adviser*)
Dr P.D. Minor, Head, Division of Virology, National Institute for Biological Standards and Control, Potters Bar, Herts., England (*Temporary Adviser*)

Dr M. Nübling, Paul Ehrlich Institute, Langen, Germany (*Temporary Adviser*)
Dr J. Wood, Division of Virology, National Institute for Biological Standards and Control, Potters Bar, Herts., England (*Temporary Adviser*)
Dr D. Wood, Coordinator, Quality Assurance and Safety of Biologicals, World Health Organization, Geneva, Switzerland (*Secretary*)

Requirements for the use of animal cells as in vitro substrates for the production of biologicals

The WHO requirements for the use of animal cells as *in vitro* substrates for the production of biologicals (WHO Technical Report Series No. 878, 1998, Annex 1) provide, inter alia, information about a WHO cell bank of Vero cells. These cells were developed in 1987 and designated as a Master Cell Bank in 1998. Cultures of the cells are available to manufacturers and national control authorities. As at its fifty-third meeting (WHO Technical Report Series, No. xxx, in press), the Committee had been informed of possible deficiencies in the records relating to the cell bank that might have regulatory implications for the establishment of master cell banks, and a revision of the Requirements was therefore proposed. A draft amendment to the section "General considerations — continuous-cell-line substrates" in the Requirements had been prepared (WHO/BS/03.1970). The Committee adopted the draft text as the Addendum 2003 to the requirements for the use of animal cells as in vitro substrates for the production of biologicals and agreed that it should be annexed to its report (Annex 4).

Annex 4

Requirements for the use of animal cells as in vitro substrates for the production of biologicals (Addendum 2003)

Introduction

World Health Organization (WHO) Requirements for the use of animal cells as in vitro substrates for the production of biologicals (*1*) provide information on a WHO cell bank of Vero cells. These cells were developed in 1987 and designated as a Master Cell Bank in 1998. Producers of biologicals and national control authorities can obtain cultures of these Vero cells (free of charge), as well as additional information, from WHO.

At its fifty-third meeting in February 2003, the Expert Committee on Biological Standardization was informed of the outcome of a meeting of a WHO Monitoring Group on Cell Banks held in Potters Bar, England, in October 2002 (*2*). The Monitoring Group noted that significant changes in regulatory expectations and technological advances had occurred in the requirements of cell bank operation and testing since the development of the WHO Vero bank 10-87 in 1987 and its designation as a Master Cell Bank in 1998. The Committee endorsed a recommendation from the Monitoring Group that the 10-87 bank should not be considered as a Master Cell Bank for direct use in manufacturing processes. Rather, the 10-87 bank should be regarded as a Cell Seed qualified by scientific analytical consensus from which Master Cell Banks may be established for thorough requalification. The Committee noted (*3*) that it would be necessary to revise the Requirements for the use of animal cells as in vitro substrates for the production of biologicals (WHO Technical Report Series, No. 878, 1998) to accommodate this change.

Manufacturers testing regimes for Master Cell Banks derived from the WHO Vero 10-87 Cell Seed will need to extend beyond the tests used to establish the WHO Vero 10-87 bank to include techniques such as product enhanced reverse transcriptase (PERT) assays (*2*). Furthermore, manufacturers should be continually aware of current developments regarding adventitious agents and ensure that data from safety testing on banks of cells used in manufacturing processes are regularly reviewed and updated where appropriate.

The redesignation of the WHO Vero cell bank 10-87 as a Cell Seed may lead to investigations into the use of cells for manufacturing processes at higher population doublings than have previously been recommended. The potential for increased tumorigenicity at higher population doublings, among other issues, must therefore be considered (*2*). This assessment of tumorigenicity should take into account the variation that may occur in assessment of population doublings (both between and within laboratories), the potential for variability of *in vivo* tumorigenicity tests and the variation between different cell culture processes. The establishment of arbitrary passage limits for the use of cells in a manufacturing process may be less important than careful process validation and testing of cells passaged beyond the process limits.

The amendments to the 1998 Requirements apply to the general considerations section and are listed below.

General considerations

Continuous-cell-line substrates

The last paragraph on page 23 of WHO Technical Report Series No. 878 currently reads:

"The WHO master cell bank of Vero cells is stored at the European Collection of Animal Cell Cultures (ECACC), Porton Down, England and the American Type Culture Collection (ATCC), Rockville, MD, USA. Producers of biologicals and national control authorities can obtain cultures of these Vero cells (free of charge), as well as additional background information, from Biologicals, World Health Organization, 1211 Geneva 27, Switzerland."

This paragraph should be replaced by the following:

"The WHO 10-87 cell bank of Vero cells is stored at the European Collection of Animal Cell Cultures (ECACC), Porton Down, England and the American Type Culture Collection (ATCC), Rockville, MD, USA. This cell bank should be regarded as a Cell Seed qualified by scientific analytical consensus from which Master Cell Banks may be established for thorough re-qualification. Producers of biologicals and national regulatory authorities can obtain cultures of these Vero cells (free of charge), as well as additional background information, from Quality Assurance and Safety of Biologicals, World Health Organization, 1211 Geneva 27, Switzerland."

References

1. Requirements for the use of animal cells as in vitro substrates for the production of biologicals. In: *WHO Expert Committee on Biological Standardization. Forty-seventh report.* Geneva, World Health Organization, 1998, Annex 1 (WHO Technical Report Series, No. 878).
2. *Report of the WHO Monitoring Group on Cell Banks, 16-17 October 2002.*
3. *WHO Expert Committee on Biological Standardization. Fifty-second report.* Geneva, World Health Organization, 2004 (WHO Technical Report Series, No. 924, page 15).

World Health Organization

ENGLISH ONLY

FINAL

Recommendations for the evaluation of animal cell cultures as substrates for the manufacture of biological medicinal products and for the characterization of cell banks

Proposed replacement of TRS 878, Annex 1

Adopted by the 61st meeting of the WHO Expert Committee on Biological Standardization, 18 to 22 October 2010. A definitive version of this document, which will differ from this version in editorial but not scientific details, will be published in the WHO Technical Report Series.

Recommendations published by the WHO are intended to be scientific and advisory. Each of the following sections constitutes guidance for national regulatory authorities (NRAs) and for manufacturers of biological products. If a NRA so desires, these Recommendations may be adopted as definitive national requirements, or modifications may be justified and made by the NRA. It is recommended that modifications to these Recommendations be made only on condition that modifications ensure that the biological product is at least as safe and efficacious as that prepared in accordance with the recommendations set out below. The parts of each section printed in small type are comments for additional guidance intended for manufacturers and NRAs, which may benefit from those details.

Table of contents

A.2.5 Special considerations for neural cell types

A.3 Selection of source materials

A.3.1 Introduction

A.3.2 Serum and other bovine-derived materials used in cell culture media

A.3.3 Trypsin and other porcine-derived materials used for preparing cell culture

A.3.4 Medium supplements and general cell culture reagents derived from other sources used for preparing cell cultures

A.4 Certifications of cell banks by the manufacturer

A.4.1 Cell line data

A.4.2 Certification of PCCs

A.4.3 Certifications of DCLS; CCLs@aol.com; and SCLs

A.5 Cryopreservation and cell banking

A.5.1 Cryopreservation

A.5.2 Cell banking

A.5.3 WHO reference cell banks

Part B. Recommendations for the characterization of cell banks of animal cell substrates

B.1 General considerations

B.2 Identity

B.3 Stability

B.4 Sterility

B.5 Viability

B.6 Growth characteristics

B.7 Homogeneity

B.8 Tumourigenicity

B.9 Oncogenicity

1. Introduction

Cell substrates are cells used to manufacture a biological product. It is well established that cell substrates themselves and events linked to cell growth can affect the characteristics and safety of the resultant biological products. Therefore, a thorough understanding of the characteristics of the cell substrate is essential in order to identify points of concern and to develop a quality control system that addresses those points.

Recent advances in the use and quality control of new animal cell substrates, particularly continuous cell lines (CCLs) and insect cells, led to the conclusion that an update of the current WHO Requirements (TRS 878) [1] should be prepared. In order to facilitate the resolution of regulatory / scientific issues related to the use of animal cell cultures, including human, as substrates for the production of biological products, WHO initiated this revision of its Requirements on cell substrates by establishing a Study Group (SG). Animal cells refer to cells derived from organisms classified in the animal kingdom. This document is the result of the SG effort, including wide consultations with individuals and organizations with expertise in this area. After receiving comments from this consultative process, as well as from invited reviewers, further revision of the draft recommendations was undertaken and presented to the ECBS in 2010. During the development of this document, guidances on this topic issued by other relevant organizations were considered. Effort was made to be compatible with the existing guidances, whenever possible.

These recommendations provide guidance to National Regulatory Authorities (NRAs), National Control Laboratories (NCLs) and manufacturers on basic principles and, in some cases, on detailed procedures that are appropriate to consider in the characterization of animal cells that are proposed for use in the manufacture of biological products. Although the decision-making authority lies with the NRA, it is advisable that NCL experts on this topic be consulted.

2. Historical overview

Historically, the major concerns regarding the safety of biological medicinal products manufactured in animal cells have been related to the possible presence of microbial contaminants and, in some cases, to the properties and components of the cells themselves such as DNA and proteins.

For example, in 1954, an experimental adenovirus vaccine was being developed, and human tumour cells (HeLa) were rejected as the cell substrate in favour of "normal" cells [2]. At that time, relatively little was known about the biological mechanism(s) that lead to human cancer, so that the risks to the recipients of a vaccine based on HeLa cells could not be assessed and quantified scientifically. Although "normal" cells were not defined, that decision led to the use of primary cell cultures (PCCs) from animals such as monkeys, hamsters, and embryonated eggs for vaccine research and development [3].

The first requirements for cell substrates were published by WHO in 1959 for the production of inactivated poliomyelitis vaccine in PCCs derived from the kidneys of clinically healthy monkeys [4]. Those requirements were revised and published in 1966 [5]. Subsequently, other PCCs were used for the production of other viral vaccines.

In the 1960s, human diploid cells (HDCs) were developed and proposed as an alternative to primary monkey kidney cell cultures for polio virus vaccine production as well as for other viral vaccines. The rationale for using HDCs was based on the ability to: a) cryogenically preserve the cells at low population doubling levels (PDLs); b) establish and characterize cryopreserved banks of cells that later could be expanded to provide a standardized source of cells for many decades; c) extensively test recovered cells before use in vaccine production; and d) demonstrate that the cells were free from detectable adventitious agents and that they were unable to form tumors when inoculated into immunosuppressed animals. Thus, HDCs were normal by all of the then existing criteria. It was argued that because HDCs were normal and could be standardized, tested, and used for many years, they were a significant improvement over PCCs.

The pathway to acceptance of HDCs was difficult and lengthy, primarily because some members of the scientific community believed that HDCs might contain a latent and unknown human oncogenic agent, and that such a theoretical agent posed a risk to recipients of vaccines produced in HDCs. Numerous conferences and discussions of new data eventually led to the acceptance of HDCs as a substrate for viral vaccine production, and they continue to be used by many manufacturers for various viral vaccines that have a long history of safety and effectiveness. The concept of a master cell bank (MCB) and working cell bank (WCB) system and characterization of the cell substrate were introduced during that period [6,7].

Both our understanding of tumor cell biology and the technological tools that were available at that time were much more limited than they are today. As a result, the proponents of using HDCs for vaccine production based their argument that the cells were normal, and therefore safe to use, on four points: a) freedom from detectable adventitious agents; b) the finite life of HDCs; c) the diploid nature of HDCs; and d) the inability of HDCs to form tumors in various *in vivo* test systems.

In order to provide a high level of assurance that those four characteristics were stable, the initial lot-release tests for each batch of a vaccine derived from HDCs included tests of the cell substrate for adventitious agents, karyology, and tumourigenicity [8,9]. The main question that was being addressed by the routine use of tumourigenicity tests was whether or not the production cell culture had undergone a contamination or transformation event such that it contained a mixture of "normal" and tumourigenic cells. It was eventually agreed that tumourigenicity testing was not sensitive enough to detect a low level of tumourigenic cells, and that it was wasteful of animals and time in repeated testing of a cell line that had been well characterized and would be used in the context of a cell-bank system. Therefore, tumourigenicity tests eventually were required only for the characterization of a MCB (using cells at the proposed *in vitro* cell age for production or beyond) for both HDCs and CCLs [10,11].

In the 1970s, there was a clinical research need for more interferon alpha (IFNα) than could be produced from primary human lymphocytes. In response to that need, human tumour cells (Namalwa) grown *in vitro* were proposed as a cell substrate for the production of IFNα. The primary concerns about the use of Namalwa cells were that they contained the Epstein-Barr virus (EBV) genome integrated into the cellular DNA, and that either whole virus or DNA containing viral elements could be transmitted to the recipients of the IFNα product. Nevertheless, by the end of the 1970s, regulatory agencies had allowed human clinical studies to commence, and the product was eventually approved in several countries. Among the most important factors that contributed to those decisions was the fact that IFN, as opposed to live viral vaccines, was not a replicating agent, and IFNα was being used as a therapeutic rather than a prophylactic product, thus representing different risk/benefit considerations. In addition, technology had advanced significantly so that IFNα could be highly purified and the purification process could be validated to demonstrate that EBV and cellular DNA were undetectable in the final product, within the limits of the assays then available, which permitted risk mitigation.

In the 1980s, advances in science and technology led to the development of recombinant DNA (rDNA) derived proteins and monoclonal antibodies (MAbs). Animal cells with the capacity to grow continuously *in vitro* (CCLs) were the substrates of choice for those products because of the ease with which they could be transfected and engineered. Also, in contrast to PCCs and HDCs, they grew rapidly to achieve a high density and expressed a variety of products at high concentrations. Chinese hamster ovary (CHO) cell lines became widely used for rDNA products, and hybridomas of various types were required for the production of MAbs. The use of such cells as substrates in the manufacture of a large array of potentially important biological medicinal products raised safety concerns once again. A scientific consensus emerged from numerous conferences that there are three major elements of potential concern related to animal cell substrates: DNA, viruses, and transforming proteins. In 1986, WHO established a SG to examine cell-substrate issues in greater depth.

The SG concluded that there is no reason to exclude CCLs from consideration as substrates for the production of biologicals, and that CCLs are in general acceptable when the manufacturing process is shown to eliminate potential contaminating viruses pathogenic for humans and to reduce DNA to acceptable levels and/or to eliminate its biological activity [12]. The SG's emphasis on infectious agents as the major risk factor was based in large part on actual experience in which virus transmission and disease had occurred through contaminated biological products (*e.g.*, hepatitis B virus and human immunodeficiency virus (HIV) in Factor VIII). WHO Requirements for Continuous Cell Lines used for Biologicals Production were published in 1987 [13]. Based on a review of more recent data, those Requirements were revised in 1998 to raise the acceptable level of rcDNA to 10 ng per parenteral dose. In addition, it was pointed out that beta-propiolactone, a viral inactivating agent, may also destroy the biological activity of DNA. Use of this agent therefore provides an additional level of confidence even when the amount of DNA per dose may be substantial [1].

During the 1990s, and on into the 2000s, a variety of CCLs were explored as cell substrates for biological products in development because, like the cell lines already mentioned, they offered significant advantages during production (*e.g.*, rapid growth and high expression). These include the following tumourigenic cell lines: HeLa for adeno-associated virus vectored HIV vaccines; PER.C6 for influenza and HIV vaccines; Madin-Darby Canine Kidney (MDCK) for influenza vaccines; and 293ORF6 for HIV vaccines. More recently insect cell lines and stem cell lines (SCLs) have been proposed for the manufacture of biological products, and such cells introduce a new set of challenges for their evaluation and characterization.

The acceptability of a given cell type (primary, diploid, stem, or continuous) as a substrate for the production of a specific biological product depends on a variety of factors including an in depth knowledge of its basic biological characteristics. In that regard, it is important to recognize that the tumourigenic potential of a CCL is but one of many factors to consider such as the extent to which the manufacturing process reduces or eliminates cellular factors that may be of concern. An assessment of the totality of the data available is needed in order to determine whether a product manufactured in a given cell substrate is potentially approvable.

The following recommendations provide guidance to manufacturers and NRAs/NCLs on the evaluation of animal cell cultures used as substrates for the production of biological medicinal products, and for the characterization of cell banks.

The main changes from the requirements published in TRS 878, annex 1, include:

1. general manufacturing recommendations applicable to all types of cell culture production have been updated;

2. some considerations for the evaluation of new cell substrates such as insect cells and stem cells (SCs) have been added;
3. definitions have been updated and expanded in number and scope, and moved to an earlier point in the document;
4. the structure of the document has been modified to include more background information, and the applicability of various sections to different types of cell substrates is highlighted;
5. a new section on risk reduction strategies during the manufacture of biological products has been added;
6. a section on Good Cell Culture Practice has been added;
7. the section of selection of source materials has been updated, and the detailed methods used to test for bovine viruses in serum were added in Appendix 1;
8. tumourigenicity testing has been updated, and a model protocol for the nude mouse model was added in Appendix 2;
9. oncogenicity testing of tumourigenic cell lysates was added, and a model protocol was added in Appendix 3;
10. recommendations for acceptable levels of residual cellular DNA are product specific and not specifically addressed; and
11. recommendations for microbial agents testing have been updated.

3. Scope

These recommendations supersede previous WHO requirements or recommendations describing procedures for the use of animal cell substrates for the production of biological medicinal products [1, 13].

Some of these recommendations also may be useful in the quality control of specific biological products during the manufacturing process, but it is beyond the scope of this document to recommend quality control release tests. Like-wise, risk-based assessments related to product approvals are beyond the scope of this document. Requirements or recommendations for individual products should be consulted in that regard.

Cells modified by recombinant DNA technology have been increasingly used in the manufacture of novel medicinal products, and specific considerations for those products are addressed elsewhere [1,10,14,15]. Nevertheless, there are a number of generic issues that apply to genetically modified as well as to other cell substrates.

These recommendations specifically exclude all products manufactured in embryonated eggs, microbial cells (*i.e.*, bacteria and yeast), and plant cells. Also excluded are whole, viable animal cells such as SCs when they are used directly for therapy by transplantation into patients or when they are developed into SCLs for the purpose of using them as therapeutic agents by transplantation. In those cases, characterization tests should be discussed with the NRA/NCL. However, SCLs used for the production of biological products such as growth factors and vaccines should comply with these recommendations.

Some of the general recommendations given here (see sections A.1 – A.5) are applicable to all animal cell substrates. More specific guidance for PCCs can be found in the relevant documents published by WHO (*e.g.*, production of poliomyelitis vaccine in primary monkey kidney cells) [4,5].

Cell substrates should be developed and used in accordance with applicable requirements of the NRA/NCL.

In general, it is not a practice consistent with Good Manufacturing Practices to re-test materials that have already been released for further manufacture, so justification would be necessary before such re-testing is undertaken. Thus, the scope of this document is intended to cover cell substrates as new cell banks are established. Specific circumstances under which re-testing of already established and released cell banks would be appropriate should be discussed with the responsible NRA/NCL.

Recommendations published by WHO are intended to be scientific and advisory in nature. The parts of each section printed in normal type have been written in the form of recommendations so that, should a NRA/NCL so desire, they may be adopted as they stand as the basis of national or regional requirements. The parts of each section printed in small type are comments or additional points that might be considered in some cases.

4. Definitions

The definitions given below apply to the terms used in these recommendations. They may have different meanings in other contexts.

***Adventitious agent*:** Contaminating microorganisms of the cell culture or source materials including bacteria, fungi, mycoplasmas/spiroplasmas, mycobacteria, rickettsia, protozoa, parasites, transmissible spongiform encephalopathies (TSE) agents, and viruses that have been unintentionally introduced into the manufacturing process of a biological product.

The source of these contaminants may be from the legacy of the cell line, the raw materials used in the culture medium to propagate the cells (in banking, in production, or in their legacy), the environment, personnel, equipment, or elsewhere.

Biological medicinal product: Biological medicinal product is a synonym for biological product or biological described in WHO Technical Report Series. The definition of a medicinal substance, used in treatment, prevention or diagnosis, as a "biological" has been variously based on criteria related to its source, its amenability to characterization by physicochemical means alone, the requirement for biological assays, or on arbitrary systems of classification applied by regulatory authorities. For the purposes of WHO, including the present document, the list of substances considered to be biologicals is derived from their earlier definition as "substances which cannot be fully characterized by physicochemical means alone, and which therefore require the use of some form of bioassay". However, developments in the utility and applicability of physicochemical analytical methods, improved control of biological and biotechnology-based production methods, and an increased applicability of chemical synthesis to larger molecules, have made it effectively impossible to base a definition of a biological on any single criterion related to methods of analysis, source or method of production. Nevertheless, many biologicals are produced using in vitro culture systems.

Developers of such medicinal products that do not fit the definition of biological medicinal product provided in this document should consult the relevant NRAs for product classification and licensing application pathway.

Biotherapeutic: For the purpose of this document, a biotherapeutic is a biological medicinal product with the indication of treating human diseases.

***Cell bank*:** A cell bank is a collection of appropriate containers whose contents are of uniform composition stored under defined conditions. Each container represents an aliquot of a single pool of cells.

The individual containers (*e.g.*, ampoules, vials) should be representative of the pool of cells from which they are taken and should be frozen on the same day by following the same procedure and by using the same equipment and reagents.

Cell culture: The process by which cells are grown *in vitro* under defined and controlled conditions where the cells are no longer organized into tissues.

***Cell line*:** Type of cell population with defined characteristics that originates by serial subculture of a primary cell population that can be banked.

Cloning and sub-coning steps may be used to generate a cell line. The term cell line implies that cultures from it consist of lineages of some of the cells originally present in the primary culture.

Cell seed: A quantity of well-characterized cells stored frozen under defined conditions, such as in the vapour or liquid phase of liquid nitrogen in aliquots of uniform composition derived from a single tissue or cell, one or more of which would be used for the production of a master cell bank. Also referred to as a pre-MCB or seed stock. May be made under Good Manufacturing Practices (GMP) conditions or under manufacturer's research and development conditions.

Cell substrate: Cells used to manufacture a biological product.

The cells may be primary or cell lines, and may be grown in monolayer or suspension culture conditions. Examples of cell substrates include primary monkey kidney, MRC-5, CHO, and Vero cells.

Cells used to generate essential components of a final product, such as Vero cells for the generation of "reverse genetics" virus for use in seeding vaccine production, are considered to be "pre-production" cell substrates. Whereas cells used to manufacture the bulk product (*e.g.*, packaging cell lines for gene therapy vectors; Vero cells for vaccine production; CHO cells for recombinant protein expression) are considered to be "production" cell substrates.

Continuous cell line (CCL): A cell line having an apparently unlimited capacity for population doublings. Often referred to as "immortal" and previously referred to as "established".

Diploid cell line (DCL): A cell line having a finite *in vitro* lifespan in which the chromosomes are paired (euploid) and are structurally identical to those of the species from which they were derived.

While this definition is accurate for standard chromosome preparations, a given human diploid cell line may contain genetic variations that will be reflected in a Giemsa-banding pattern that differs from the standard. Gene expression differences also may be found.

This definition is based on experience and current understanding of the *in vitro* behavior of human cells that are not of stem cell origin.

DNA infectivity: The capacity of cellular DNA to generate an infectious virus following introduction of that DNA into appropriate cells. The viral genome could be integrated or extrachromosomal.

***Endogenous virus**:* A virus whose genome is present in an integrated form in a cell substrate. Endogenous viruses are present in the genome of the original animal from which the cells were derived. They may or may not encode an intact or infectious virus. -

End of production cells (EOPC): Cells harvested at or beyond the end of a production (EOP) run.

In some cases, production cells are expanded under pilot-plant scale or commercial-scale conditions.

Extended cell bank (ECB): Cells cultured from the MCB or WCB propagated to the proposed *in vitro* cell age used for production or beyond.

Functional integrity: The culture sustains the expected performance related to its intended use under specified conditions (*e.g.,* expression of secreted product at a consistent level; production of expected yield of virus).

Immortalized: having an apparently unlimited capacity for population doubling.

Indicator cells: Cells of various species used in the *in vitro* adventitious agent test that are intended to amplify adventitious viruses to promote their detection. Generally, this would include a human diploid cell line, such as MRC-5, a monkey kidney cell line, such as Vero cells, and a cell line of the same species and tissue as the cell bank. The purpose of these cell lines is to *indicate* a viral infection of the cell bank either through observation of cytopathic effect during and after an appropriate observation period or by hemadsorption and/or hemagglutination at the end of the observation period. Thus, they are referred to as *indicator* cells. The cell bank may be analyzed on such indicator cells either by co-cultivation or by passage of cell lysates or spent culture supernatants from the cell bank onto the indicator cells.

In vitro cell age: Measure of time between thaw of the MCB vial(s) to harvest of the production vessel measured by elapsed chronological time, by population doubling level of the cells, or by passage level of the cells when subcultivated by a defined procedure for dilution of the culture.

Latent virus: A virus is considered to be latent when the viral genome is present in the cell without evidence of active replication, but has the potential to be reactivated.

***Master cell bank (MCB)*:** A quantity of well-characterized cells of animal or other origin, derived from a cell seed at a specific population doubling level (PDL) or passage level, dispensed into multiple containers, cryopreserved, and stored frozen under defined conditions, such as the vapour or liquid phase of liquid nitrogen in aliquots of uniform composition. The master cell bank is prepared from a single homogeneously mixed pool of cells. In some cases, such as genetically engineered cells, this may be prepared from a selected cell clone established under defined conditions. Oftentimes, however, the MCB is not clonal. It is considered best practice that the master cell bank is used to derive working cell banks.

Oncogenicity: The capacity of an acellular agent such as a chemical, virus, viral nucleic acid, viral gene(s), or a subcellular element(s) to cause normal cells of an animal to form tumours.

Oncogenicity is distinct from tumourigenicity (see Tumourigencity). The tumours that arise in an oncogenicity test are of host origin whereas in a tumorigenicity test, the tumors are derived from the inoculated cells.

Parental cells: Cells that are manipulated to give rise to a cell substrate. For hybridomas, it is typical to also describe the parental cells as the cells to be fused.

Manipulation may be as simple as the expansion of a primary cell culture to provide early passage cells, or a more complex activity such as developing a hybridoma or transfected clone, and both processes would provide a cell seed. The parental cells may refer to any stage prior to the preparation of the cell seed. Examples of a parental cell are: WI-38 and MRC-5 at very early passage; Vero at passage 121; and CHO before the introduction of a DNA construct to produce a recombinant cell. In certain situations (*e.g.*, myeloma cells), there may be a lineage of identified stable parental clones, thus, the term "parental cell" would normally refer to the cells used immediately prior to generation of the "cell seed".

Passage: Transfer of cells, with or without dilution, from one culture vessel to another in order to propagate them, and which is repeated to provide sufficient cells for the production process.

This term is synonymous with "subculture". Cultures of the same cell line with the same number of passages in different laboratories are not necessarily equivalent because of differences in cell culture media, split ratios, and other variables that may affect the cells. This is a more important consideration for SCLs and CCLs than for DCLs. Population doubling is the preferred method of estimating cell line age, and whenever possible, it should be used instead of "passage". However, it also may be appropriate to quantify culture duration of CCLs by the number of subcultivations at a defined seeding density at each passage or time in days.

Population doubling: A two-fold increase in cell number.

Population doubling level (PDL): The total number of population doublings of a cell line or strain since its initiation *in vitro*. A formula to use for the calculation of "population doublings" in a single passage is: number of population doublings = $Log_{10}(N/N_o) \times 3.33$ where: N = number of cells in the growth vessel at the end of a period of growth. N_o = number of cells plated in the growth vessel [16]. It is best to use the number of viable cells or number of attached cells for this determination.

Primary culture: A culture started from cells, tissues or organs taken directly from one or more organisms. A primary culture may be regarded as such until it is successfully subcultured for the first time. It then becomes a "cell line" if it can continue to be subcultured at least several times.

Production cell cultures: A collection of cell cultures used for biological production that have been prepared together from one or more containers from the working cell bank or, in the case of primary cell cultures, from the tissues of one or more animals.

Residual cellular DNA (rcDNA): cell substrate DNA present in the final product.

Specific pathogen free (SPF): Animals known to be free of specific pathogenic microorganisms and reared in an environment that maintains that state. SPF animals usually are raised in biosecure facilities, and their health status is monitored on an ongoing basis. The SPF status simply provides an assurance that the stock is not infected with the specified pathogens. SPF animals are not disease free nor are they disease resistant. They may carry pathogens other than those from which they are specified to be free.

Stem cell line: A continuous cell line generated from stem cells, rather than normal or diseased differentiated tissue.

Transmissible Spongiform Encephalopathy (TSE): The transmissible spongiform encephalopathies (TSEs) are a group of fatal neurodegenerative diseases which include classical and variant Creutzfeldt–Jakob disease (CJD), Gerstmann-Sträussler-Scheinker syndrome (GSS), fatal familial insomnia (FFI), and Kuru in humans, bovine spongiform encephalopathy (BSE) in cattle, chronic wasting disease (CWD) in mule deer and elk, and scrapie in sheep and goats.

Tumourigenicity: The capacity of a cell population inoculated into an animal model to produce a tumour by proliferation at the site of inoculation and/or at a distant site by metastasis.

Tumourigenicity is distinct from Oncogenicity (see Oncogenicity).

WHO Reference cell bank: A cryopreserved stock of cells prepared from a single, homogeneous pool of cells prepared under defined conditions and subjected to characterization tests. The purpose of such a bank is to serve as a well-characterized cell seed for the preparation of master cell banks that will be extensively characterized by manufacturers, and that has a high probability of meeting these recommendations.

***Working cell bank (WCB)*:** A quantity of well-characterized cells of animal or other origin, derived from the master cell bank at a specific PDL or passage level, dispensed into multiple containers, cryopreserved, and stored frozen under defined conditions, such as in the vapour or liquid phase of liquid nitrogen in aliquots of uniform composition. The working cell bank is prepared from a single homogeneously mixed pool of cells. One or more of the WCB containers is used for each production culture.

5. General considerations

Types of animal cell substrates

Primary Cell Cultures (PCCs)

PCCs have played a prominent role in the development of biology as a science, and of virology in particular. Cultures of PCCs from different sources have been in worldwide use for the production of live and inactivated viral vaccines for human use for many decades. For example, PCCs of monkey kidney cells have been used for the production of inactivated and oral poliomyelitis vaccines since the 1950s.

Major successes in the control of viral diseases, such as poliomyelitis, measles, mumps and rubella, were made possible through the wide use of vaccines prepared in PCCs, including those from chicken embryos and the kidneys of monkeys, dogs, rabbits and hamsters, as well as other tissues.

PCCs are viable cells of disaggregated tissues that are initiated as *in vitro* cell cultures usually as adherent cells. Many cell types will be present and a primary culture may be a complex mixture of cells that can be influenced by the process and conditions under which they were harvested, disaggregated and introduced to *in vitro* culture. Not all cells in a primary culture will have the capacity to replicate. Particular care should be applied to establishing highly reproducible procedures for tissue disaggregation, cell processing and culture initiation and reproducible culture conditions and nutrition.

PCCs obtained from wild animals usually show a high frequency of viral contamination. For example, monkey-kidney cell cultures can be contaminated with one or more adventitious agents, including simian viruses.

If PCCs are necessary for the production of a given biological, then the frequency of contaminated cell cultures can be significantly reduced by careful screening of the source animals for the absence of such viruses. Viruses can be detected by molecular tests such as PCR, and by looking for the presence of circulating antibodies to those viruses in the source animals. The use of animals bred in a carefully controlled colony, especially those that are specific-pathogen-free, is strongly recommended. Nevertheless, as suitable alternative cell substrates become available, PCCs are less likely to be used in the future. WHO has promoted the replacement of animals for experimental purposes both from an ethical perspective [17] and in the interests of progressive improvement in product safety and quality.

Advantages: (a) they are comparatively easy to prepare using simple media and bovine serum; and (b) they generally possess a broad sensitivity to various viruses, some of which are cytopathic.

Disadvantages: (a) contamination by infectious agents is a higher risk than with DCLs and CCLs; (b) the quality and viral sensitivity of cultures obtained from different animals are variable; and (c) although cell cultures derived from nonhuman primates had been in wide use in the past, it has become increasingly difficult to obtain and justify the use of such animals for this purpose, (d) they cannot be tested as extensively as DCLs or CCLs.

Diploid Cell Lines (DCLs)

The practicality of using human DCLs for the production of viral vaccines was demonstrated in the 1960s. The experience gained with oral poliomyelitis and other viral vaccines in successfully immunizing billions of children in many countries has shown clearly that such substrates can be used in the production of safe and effective vaccines [3].

The essential features of DCLs of human (*e.g.*, WI-38, MRC-5) and monkey (*i.e.*, FRhL-2) origin are: (a) they are cells passaged from primary cultures that have become established as cell lines with apparently stable characteristics over numerous PDLs; (b) they have a finite capacity for serial propagation, which ends in senescence, a state in which the culture ceases to replicate, the cells remain alive and metabolically active, but may show morphological and biochemical changes some of which begin to appear before replication ceases; (c) they are non-tumourigenic; and (d) they display diploid cytogenetic characteristics with a low frequency of chromosomal abnormalities of number and structure until they enter senescence. Substantial experience since the 1960s has been accumulated on the cytogenetics of WI-38 and MRC-5, and ranges of expected frequencies of chromosomal abnormalities have been published [18,19]. Similar data are available for FRhL-2 [20]. More sophisticated cytogenetic techniques (*e.g.*, high resolution banding, comparative genome hybridization) [21,22] have demonstrated the presence of subtle chromosomal abnormalities that were previously undetectable. Recent studies have shown that subpopulations of human DCLs with such abnormalities may appear and disappear over time, and that they are non-tumourigenic and undergo senescence in the same manner as the dominant population. Thus, possessing a stable karyotype might not be such an important characteristic as was previously thought.

Advantages: (a) they can be well characterized and standardized; (b) production can be based on a cryopreserved cell-bank system that allows for consistency and reproducibility of the reconstituted cell populations. A cell-bank system usually consists of cell-banks of defined population doubling or passage levels that generally include a MCB and a working cell-bank (WCB); and (d) unlike the CCLs and SCLs discussed below, DCLs are not tumourigenic and therefore do not raise the potential safety issues associated with CCLs and SCLs.

Disadvantages: (a) they are not easy to use in large-scale production, although they have been cultivated using bioreactor technology employing the microcarrier or multilayer method; (b) in general, they have more fastidious nutritional requirements than other cell substrates; (c) they may be difficult to adapt to serum-free growth; and (d) they are more difficult than CCLs to transfect and engineer, and require immortalization before they can be engineered; (*e.g.*, they are not permissive for the production of vaccine vectors that require complementation, since they cannot be engineered readily to express complementing proteins).

Continuous cell lines (CCLs)

Some CCLs have been used for the production of safe and effective biotherapeutics and vaccines since the 1980s.

CCLs have the potential for an apparently indefinite *in vitro* life span and have been derived by the following methods: (a) serial subcultivation of a PCC of a human or animal tumour (*e.g.*, HeLa cells); (b) transformation of a normal cell having a finite life span with an oncogenic virus or viral sequence (*e.g.*, B lymphocytes transformed by EBV or transfected with viral sequences such as in PER.C6); (c) serial subcultivation of a primary-cell population derived from normal tissue that generates a dominant cell population having an apparently indefinite life span, often described as spontaneous transformation (*e.g.*, Vero, BHK-21, CHO, MDCK, Hi5); (d) fusion between a myeloma cell and an antibody-producing B lymphocyte to produce a hybridoma cell line; or (e) use of ectopically expressed genes involved in the cell cycle such as hTERT telomerase gene to enable indefinite replication of normal human cells.

CCLs may display a consistent modal chromosome number (*e.g.*, MDCK, Vero), and although the karyotype of individual cells in a culture at any one time-point may vary, the range of chromosome numbers per cell will usually show characteristic limits.

However, other CCLs, such as highly tumourigenic cells including HeLa may show variation in modal number and a wider drift in the range of the number of chromosomes per cell.

In the early stage of establishing a cell line, significant diverse karyotypes and changes in karyotype may be observed, but a characteristic typical chromosome component may emerge with continued passage presumably as a dominant cell population develops.

Advantages: (a) they can be characterized extensively and their culture conditions standardized; (b) production can be based on a cell-bank system, which allows consistency and reproducibility of the reconstituted cell populations for an indefinite period; (c) as a rule, they grow more easily than DCLs using standard media, (d) most can be adapted to grow in serum-free medium; (e) they usually can be grown on microcarriers for large-scale production in bioreactors; and (f) some can be adapted to grow in suspension cultures for large-scale production in bioreactors.

Disadvantages: (a) CCLs may express endogenous viruses, and some are tumourigenic in immunosuppressed animal models; (b) theoretical risks identified by the 1986 Study Group (*e.g.*, nucleic acids, transforming proteins, and viruses) need to be taken into account.

Stem Cell Lines (SCLs)

SCs differ from other types of cells because they sustain a predominant stem-cell population whilst simultaneously retaining the capacity to produce cell progenitors of differentiated cell types of almost all human tissues (*i.e.*, pluripotent). Pluripotent SCLs have an apparent capacity to generate cell types of all three human germlayers and may be capable of generating *in vitro* models of any tissue in the human body. At the time these recommendations are written (2010), two types of pluripotent SCLs, human embryonic stem cells and induced pluripotency stem cell lines have been isolated which may have the capabilities to prove useful for manufacturing biologicals. This property of pluripotency is sustained through numerous cycles of cell division. SCLs may be derived from early stage embryonic, fetal or adult tissues. Typically, specialized media and environmental conditions such as the attachment matrix are required for the growth of SCLs *in vitro* in order for them to maintain the undifferentiated state. While most stem cell research and development has been directed towards transplantation of SCs for therapeutic purposes, efforts also have gone into exploring a variety of SCLs as cell substrates for the production of biologicals.

Key considerations for the culture and control of such cell lines have been developed [23]. These include the fundamental issues common to the maintenance of all cell lines, but also emphasize the need for appropriate ethical governance regarding donor consent and careful attention to periodic confirmation of phenotype, absence of non-diploid cells, and sustained pluripotent capacity.

Recently, it has been shown that conditioned medium from SCLs can have regenerative properties, and such preparations produced from human embryonic SCs have shown regenerative capabilities including repair of myocardial infarction in animal models [24]. This raises the possibility of SCs being used as a substrate to produce a variety of biologically active molecules. SCLs can, in some respects, be considered as diploid cells, but they do not appear to have the finite life span characteristic of human diploid fibroblast cultures. In human embryonic stem-cell cultures, clonal variants with chromosomal abnormalities are known to arise. Whilst a diploid and non-transformed nature is to be considered a pre-requisite for cell therapy applications, transformed SCLs might be considered as a form of CCL for the manufacture of biologicals. Because they do not fall easily within any one category of substrate already discussed, SCLs are identified separately in this document.

Advantages

(a) they can be well characterized and their culture conditions standardized; (b) production can be based on a cell-bank system, which allows consistency and reproducibility of the reconstituted cell populations for an indefinite period; and (c) some may be adapted to grow in suspension cultures for large-scale production in bioreactors; (d) they may produce unique proteins of potential importance as biotherapeutics; and (e) they have the potential to generate cells and tissue-like structures that may permit the expression of agents currently considered "unculturable" *in vitro*.

Disadvantages

(a) Subculture techniques commonly used for SCLs are laborious; (b) may produce growth proteins with undefined effects on adult cells/tissues; (c) usually require complex media that may have a TSE risk; (d) rapid development of differentiated cells also means that they are difficult to control *in vitro*; (e) there is little experience with their use as a cell substrate to manufacture biological products.

Potential risks and risk mitigation associated with biologicals produced in animal cell cultures

The main potential risks associated with the use of biologicals produced in animal cells are directly related to contaminants from the cells, and they fall into three categories: (a) viruses and other transmissible agents; (b) cellular nucleic acids (DNA and RNA); and (c) growth-promoting proteins. In addition, cell-derived inhibiting or toxic substances are theoretically possible. A summary of the risk assessment for each follows. More comprehensive information has been published elsewhere on the risks associated with contaminating DNA and growth-promoting proteins [25,26,27,28,29,30,31,32,33,34].

Early in 2010 NRAs as well as WHO were made aware of new information regarding the presence of DNA sequences of porcine circovirus in live attenuated rotavirus vaccines. The detection of these sequences by the use of advanced analytical methods raised complex questions, *e.g.*, the evaluation of the potential risk, specific testing of vaccines and the general use of these methods for the characterization of vaccine cell substrates. The power of the new methodology that was used (*i.e.*, massively parallel (deep) sequencing) may uncover the presence of adventitious agents that might not be detected with current methods. While the implementation for routine use of such methods has benefits as well as challenges and risks, NRAs need to be prepared for similar situations. Considerations for making a risk assessment and potentially introducing risk mitigation strategies may need to be undertaken in such circumstances.

Viruses and other transmissible agents

There is a long history of concern regarding the potential transmission of viruses and other infectious agents that may be present in cell substrates. This area was reviewed most recently by WHO in 1986 by a SG who pointed out that, as described below, cells differ with respect to their potential for carrying viral agents pathogenic for humans.

Primary monkey kidney cells have been used to produce billions of doses of poliomyelitis vaccines since they were first developed in the 1950s, and although viruses such as SV40 were discovered in rhesus monkey kidney cells, control measures were introduced to eliminate or reduce as much as possible the risk of viral contamination associated with the manufacture of vaccines in cells containing those viruses. Additional controls may be needed as new viral agents are identified and technologies to detect them are developed.

Human and nonhuman primate lymphocytes and macrophages may carry latent viruses, such as herpesviruses and retroviruses. CCLs of non-haematogenous cells from human and nonhuman primates may contain viruses or have viral genes integrated into their DNA. In either case, virus expression may occur under *in vitro* culture conditions.

Avian tissues and cells may harbour exogenous and endogenous retroviruses, but there is no evidence for transmission of disease to humans from products prepared using these substrates. For example, large quantities of yellow fever vaccines were produced for many years in eggs that contain avian leukosis viruses, but there is no evidence that these products have transmitted disease in their long history of use for human immunization. Nevertheless, the potential for transmitting avian retroviruses should be reduced as much as possible through manufacturing control measures.

Rodents may harbour exogenous and endogenous retroviruses, lymphocytic choriomeningitis virus, and Hantaviruses, and a range of other potentially zoonotic viruses. While contamination with the rodent viruses in the cell harvests of biotherapeutic products derived from CHO cell culture has been reported [34,35,36], there is no evidence that biological products released for distribution have been contaminated with rodent viruses, because, if present, they were detected during quality control testing in compliance with GMPs prior to release. In addition, it is important to note that there have been no reported cases of transmission of an infectious agent to recipients of recombinant protein products manufactured in animal cells.

Insect cells recently have been used for vaccine production, and various insect cell lines may be used for the production of biologicals in the future. Insect viruses tend to be ubiquitous in many insect cell lines, and are generally unknown and/or uncharacterized. Many insect cell lines have endogenous transposons and retrovirus-like particles, and some are positive in PERT assays.

HDCs have been used for vaccine production for many years, and although concern was initially expressed about the possibility of such cells containing a latent pathogenic human virus, no evidence for such an endogenous agent has been reported, and vaccines produced from this class of cell substrate have proven to be free from viral contaminants.

In light of the differing potential of the various types of cells mentioned above for transmitting viruses that are pathogenic in humans, it is essential that the cells being used to produce biological products be evaluated as thoroughly as possible with respect to infectious agents.

When DCLs, SCLs, or CCLs are used for production, a cell bank system should be used and the cell banks should be characterized as specified in this document. Efforts to identify viruses by testing for viral sequences or other viral markers, especially those not detectable by other means, constitute an important part of the evaluation of cell banks in addition to the standard tests that have been in place for many years.

When cell lines of rodent or avian origin are examined for the presence of viruses, the major emphasis in risk assessment should be placed on the results of studies in which transmission to target cells or animals is attempted. Risk to human recipients should not be assessed solely on ultrastructural or biochemical/biophysical evidence of the presence of viral or viral-like agents in the cells.

The overall manufacturing process, including the selection and testing of cells and source materials, any purification procedures used, and tests on intermediate or final products, should ensure the absence of detectable infectious agents in the final product. When appropriate, validation of purification procedures should demonstrate adequate reduction of relevant model viruses with a significant safety factor [14]. This is usually required for recombinant protein products.

There may be as yet undiscovered microbial agents for which there is no current evidence or means of detection. As such agents become identified, it will be important to consider whether to re-examine cell banks for their presence. In general, it is not a practice consistent with GMPs to re-test materials that have already been released, so justification would be necessary before such re-testing is undertaken. Positive findings should be discussed with the NRA/NCL. Whenever new data are developed with the potential for an impact on the quality, safety or efficacy of a biological product, it is the responsibility of the manufacturer to provide NRAs all relevant data and information currently available. This should include confirmation and evaluation of the finding, the manufacturer's own risk assessment as well as an investigational and action plan, in order to facilitate any regulatory action that might be necessary. In addition, new testing methods are likely to be developed, and as they become available and validated, they should be considered by manufacturers and NRAs/NCLs for their applicability to the characterization and control of new animal cell substrates.

Cellular DNA

The issue of rcDNA in biological products has been considered by many groups since the 1980s, and there has been an evolution of consensus on recommendations during that period. The most recent WHO recommendation (TRS 878) sets the upper limit of rcDNA at 10 ng per parenteral dose. As stated below, while this value has proved helpful in the past, it does not take into consideration important factors such as the size of the DNA fragments and any potentially inactivating steps in the manufacturing process. Therefore, the amount of rcDNA that might be acceptable for a specific product should take into consideration not only the limit of 10 ng per parenteral dose, but other factors as well when determining the acceptable level of rcDNA.

PCCs and DCLs have been used successfully for many years for the production of viral vaccines, and the rcDNA deriving from these cells has not been (and is not) considered to pose any significant risk. However, with the use of CCLs, which have an apparently indefinite life span, presumably due to the dysregulation of genes that control growth, and with the ongoing development of products from cells that are tumourigenic or were derived from tumors, the DNA from such cells has been considered to have the theoretical potential to confer the capacity for unregulated cell growth, and perhaps oncogenic activity, upon some cells of a recipient of the biological product. Although the risk of such DNA has been estimated based on certain assumptions and some experimental data, assessing the actual risk of such DNA has not been possible until recently when preliminary data generated from new experimental systems began to quantify those risks [37].

The potential risks of DNA arise from both of its biological activities: (a) infectivity, and (b) oncogenicity. Infectivity could be due to the presence of an infectious viral genome in the cellular DNA of the cell substrate [38,39,40]. The viral genome could be that of a DNA virus, whether integrated or extrachromosomal, or that of a proviral genome of a retrovirus. Both types of viral DNA have been shown to be infectious *in vitro* and in several cases, *in vivo* [37,38,39]. The oncogenic activity of DNA could arise through its capacity to induce a normal cell to become transformed and perhaps to become tumourigenic. The major mechanism through which this could occur would be through the introduction of an active dominant oncogene (*e.g.*, *myc,* activated *ras*), since such dominant oncogenes could directly transform a normal cell. Other mechanisms require that the rcDNA transform through insertional mutagenesis, and have been considered less likely, since the frequency of integration of DNA in general is low [40]. The frequency of integration at an appropriate site, such as inactivating a tumor suppressor gene or activating a proto-oncogene, would be correspondingly lower [31].

The 1986 WHO SG addressed the risk posed by the oncogenic activity of rcDNA in biological products for human use [12]. Risk assessment based on a viral oncogene in an animal model suggested that *in vivo* exposure to one nanogram (ng) of rcDNA, where 100 copies of an activated oncogene were present in the genome, would give rise to a transformational event once in 10^9 recipients [26]. On the basis of this and other evidence available at that time, the 1986 SG concluded that the risk associated with rcDNA in a product is negligible when the amount of such DNA is 100 picograms (pg) or less per parenteral dose. Based on a review of more recent data, those requirements were revised in 1998 to raise the acceptable level of rcDNA to 10 ng per dose.

Studies in mice using cloned cellular oncogenes also suggest that the risk of neoplastic transformation by cellular DNA is probably very low [33,42]. However, more recent data have shown that cloned cellular oncogene DNA can induce tumors in selected strains of mice at levels below 1 ng. In addition, single oncogenes can also be biologically active [42] and initiate the tumour induction process. Because of these data, and the recent description that genes encoding for certain micro-RNA species can be oncogenic *in vitro* [44,45,46,47], thus increasing the number of potential dominant cellular oncogenes, the oncogenic risk of DNA needs to be considered when tumourigenic cells are considered for use in the production of biologicals. This would be especially important for live, attenuated viral vaccines where chemical inactivation of the DNA is not possible, and the only way the biological activity of DNA could be reduced would be by nuclease digestion and the reduction in the quantity of DNA.

In addition to its oncogenic activity, the infectivity of DNA should be considered. Since a viral genome once introduced could amplify and produce many infectious particles, the infectivity risk is likely greater than the oncogenic risk. The polyoma virus genome is infectious in mice at about 50 pg [48], and a recent report demonstrated that 1 pg of a proviral copy of a retrovirus is infectious *in vitro* [49]. Because such low levels of DNA may be biologically active, the amounts of rcDNA should be factored into safety evaluations when tumourigenic cell substrates are used, especially for live viral vaccines.

Therefore, considerations that need to be taken into account with respect to rcDNA are: (a) any reduction in the amount of the contaminating DNA during the manufacturing process; (b) any size reduction of the contaminating DNA during the manufacturing process; and (c) any chemical inactivation of the biological activity of contaminating DNA during the manufacturing process. A product might be considered by a NRA/NCL to have an acceptable level of risk associated with the DNA of the cell substrate on the basis of (a) and/or (b) and/or (c), when data demonstrate that appropriate levels have been achieved. For example, data have shown that nuclease digestion of DNA or chemical inactivation of DNA with beta-propiolactone, a viral inactivating agent, can destroy the biological activity of DNA [37,49,50]. Therefore, the use of these procedures may provide an additional level of confidence with respect to DNA risk reduction.

For products such as monoclonal antibodies and subunit vaccines manufactured in tumourigenic cell substrates, it is necessary to demonstrate the clearance (removal and/or inactivation) of DNA by the manufacturing process, which may require the validation of the main inactivating or removal steps. For example, data should be obtained on the effects of DNA-inactivating agents under specific manufacturing conditions so that firm conclusions on their DNA-inactivating potential for a given product can be drawn.

There may be instances where CCL DNA is considered to pose a higher level of risk because it contains specific elements such as infectious retroviral proviral sequences. Under these circumstances, the steps taken to reduce the risks of rcDNA, such as reducing the size of DNA fragments, should be set in consultation with the NRA/NCL.

The 1986 WHO SG stated that the risks for rcDNA should be considered negligible for preparations given orally. This conclusion was based on the finding that polyoma virus DNA was not infectious when administered orally [48]. For such products, the principal requirement is the elimination of potentially contaminating viruses. Recently, additional data on the uptake of DNA *via* the oral route have been published [51]. These studies demonstrated that the efficiency of uptake of DNA introduced orally was significantly lower than that introduced intra-muscularly. Nevertheless, the specifics of the manufacturing process and the formulation of a given product should be considered by the NRA/NCL.

With respect to the efficiency of DNA uptake *via* the nasal route, no data have been published comparing the nasal route with parenteral routes. However, data presented publically show that uptake *via* the intranasal route is less efficient than by the intramuscular route. Limits for a specific product should be set in consultation with the NRA/NCL.

In general, acceptable limits of rcDNA for specific products should be set in consultation with the NRA/NCL taking into consideration the characteristics of the cell substrate, the intended use of the biological product, and most importantly the effect of the manufacturing process on the size, quantity and biological activity of rcDNA fragments. In general, it has been possible to reduce rcDNA in biotechnological products to <10 ng per dose, and in many cases <10 pg per dose, because they can be highly purified. Quantitative PCR for short amplicons has been used to determine total residual DNA levels as well as residual DNA fragment size distribution. It should be noted that other methods may give different results for small fragments or for DNA that has been treated with inactivating agents. Whatever methods are used should be validated. Some products, especially certain live viral vaccines, are difficult to purify without a significant loss in potency, so that the amount of rcDNA in those final products may be significantly higher than 10 ng per dose. Such cases are considered to be exceptional and should be discussed with the NRA/NCL.

Cellular RNA

While protein-coding RNA has not been considered to be a risk factor for biological products due to the unstable nature of RNA and the lack of mechanisms for self-replication, the recent description of small non-coding RNA molecules – microRNA (miRNA) – that are more stable and have the capacity to modulate gene expression might necessitate a reassessment. Whether these miRNA molecules can be taken up by cells *in vivo* is unknown. However, as stated above, because certain miRNA genes can be oncogenic, DNA containing such sequences may need to be considered along with oncogenes when assessing the risk of rcDNA (see B.9 Oncogenicity). However, because this is an evolving area of research, no conclusions can be made regarding the risk of miRNA, and no recommendations are made to control miRNA at this time.

Growth-promoting proteins

Growth factors may be secreted by cells used to produce biologicals, but the risks from these substances are limited, since their growth-promoting effects are usually transient and reversible, they do not replicate, and many of them are rapidly inactivated *in vivo*. In exceptional circumstances, growth factors can contribute to oncogenesis, but even in these cases, the tumours apparently remain dependent upon continued administration of the growth factor. Therefore, the presence of known growth-factor contaminants at ordinary concentrations does not constitute a significant risk in the preparation of biological products manufactured in animal-cell cultures. However, some SCLs may secrete higher levels and more potent factors than CCLs. This should be taken into account when designing characterization studies, and the manufacturing process should be designed to address any safety issues that are identified.

Part A. General recommendations applicable to all types of cell culture production

A.1 Good manufacturing practices

The general principles of GMP for biologicals should be in place. Requirements or recommendations have been made by NRAs (*e.g.*, EMA, FDA) and other groups (*e.g.*, ICH). GMP should be applied from the stage of cell banking.

In the preparation of a cell substrate, it is considered best practice to establish tiered master and working cell banks (A.5.3) to ensure a reliable and consistent supply of cells that can be fully characterized and safety tested prior to use for production. By definition, primary-cell cultures cannot be subjected to such banking regimens. However, some manufacturers have utilized pooled and cryopreserved primary cultures, which enable completion of lot-release testing as in a tiered banking system. The strategy for delivery of primary cells or primary cells recovered from cryopreservation should be based on the quality and safety that can be assured for the final product according to the overall manufacturing and control processes involved.

A.2 Principles of Good Cell Culture Practices

A.2.1 *Understanding the cells and the culture system*

In all aspects of sourcing, banking and preparing cell cultures, the principles of Good Cell Culture Practices (GCCPs) should be observed (for example, 52, 53 for SCLs). Fundamental features to be considered in the development of cell cultures for production or testing are:

1. authenticity, including identity, provenance, and genotypic/phenotypic characteristics
2. absence of contamination with another cell line
3. absence of microbiological contamination, and
4. stability and functional integrity on extended *in vitro* passage

An important basic principle for all types of cells is that the donor should be free of transmissible diseases or diseases of uncertain etiology, such as CJD for humans and BSE for cattle. The NRA/NCL may allow specific exceptions concerning donor health (*e.g.*, myeloma, other tumour cells).

Cells in culture may change their characteristics in response to changes in culture conditions or on extended passage under the same culture conditions. The four cell culture types (PCC, DCL, SCL, CCL) used in manufacture differ in their potential stability, and characterization approaches may need to be adapted in reflection of these differences.

Cell cultures grow in an *in vitro* environment that is substantially different from the conditions experienced by cells *in vivo*, and it is not unexpected that they may be susceptible to change or alteration as a result of *in vitro* culture and processing. It is important to be conscious of the variation that may arise in the cell-culture environment, as cells may undergo subtle alterations in their cell biology in response to such changes. It is therefore necessary to try to control key known variables that could have significant impact on the cell culture. Medium and specific additives (serum, growth factors, amino acids and other growth promoting compounds) should, where possible, be specified in terms of chemical composition and purity. Where relevant, biological activity of the medium and additives should be determined before use. New batches of reagents for cell culture should be supplied with certificates of analysis and origin, which enable their suitability to be evaluated against the established specification. The use of serum or other poorly defined reagents is not recommended in the production of new biologicals from cell culture, and wherever possible chemically defined alternatives should be sought. However, given that our current understanding of cell biology is not complete, there is a balance to be made between the benefits that defined media bring in the form of higher reproducibility and reduced risk, and the potential effects of inadequacies of defined-culture systems that may not meet the full biological needs of cells. Where complex biological reagents, such as FBS remain necessary, they should be carefully controlled, whenever possible, by pre-use selection of batches. Such careful selection also should apply to cell-culture surfaces using specified culture vessels or surface coatings where relevant.

Variation in physical-culture parameters, such as pH, temperature, humidity, and gas composition, can significantly influence the performance and viability of cells and should be specified with established tolerances and the relevant equipment calibrated and monitored. In addition, any culture reagents prepared in the laboratory should be documented, quality controlled and released against an established specification.

A.2.2 *Manipulation of cell cultures*

In vitro processing of cells can introduce additional physical and biochemical stresses that could have an influence on the quality of the final product. Care should be taken to minimize manipulations, taking into consideration the specifics of the manufacturing process. In all cases, a consistent process should be demonstrated.

A.2.2.1 *Detachment and subculture*

Detachment solutions may adversely affect the cells if exposure is not minimized. Cell harvesting and passaging procedures should be carried out in a reproducible way ensuring consistency in the confluency of cells when harvested, incubation times, temperature, centrifugation speeds and times, and post-passage viable cell seeding densities.

A.2.2.2 *Cryopreservation* (see A.5.1)

A.2.2.3 *Introduction of contamination*

As already mentioned, the microbiological status of the donor individual, colony, herd, or flock is an important consideration in the establishment of PCCs. In order to avoid catastrophic failure of the production process and to avoid infectious hazards for recipients of products, it is important to minimize the opportunities for contamination of cell cultures. Therefore, cell manipulation and open processing steps should be minimized, taking into consideration the specifics of the manufacturing process. It is critical to adopt rigorous aseptic technique and provide appropriate environmental controls and air quality for cell-culture processing and preparation of growth media. The presence of any antimicrobial in a biological process or product is discouraged, although a notable exception is that antibiotic(s) and antifungal(s) may be required for primary-cell cultures. Additionally, antibiotics may be used in some cell-line selection systems. Where antibiotics have been used, sterility testing procedures should account for potential inhibitory effects of the antibiotic on contaminating organisms. Penicillin or other beta-lactam antibiotics should not be present in production-cell cultures.

A.2.3 *Training and staff*

Training in all cell-culture processes is vital to ensure correct procedures are adhered to under GMPs. Staff should be trained in the underlying principles of cell-culture procedures to give them an understanding of cell-culture processes that will enable them to identify events and changes that could impact on the quality of cells [52]. Key procedures on which such GCCPs training should focus include passaging of cells, preparation of sterile media, maintenance and use of biological safety cabinets, incubators, and cryopreservation.

Cell cultures should be prepared by staff who have not, on the same working day, handled animals or infectious microorganisms. Furthermore, cell cultures should not be prepared by staff who are known to be suffering from a transmissible infection. The personnel concerned should undergo a “return to work” assessment to evaluate any residual risk.

A.2.4 *Cell line development and cloning*

Wherever a cell culture has passed through a process that may have a significant influence on its characteristics, such as tumourigenicity, it should be treated as a new (*i.e.*, different) cell line and should be renamed with a suffix or code to identify this. A MCB should then be prepared from the post 'treatment' culture. Treatments that may require such rebanking include cell cloning and genetic manipulation. Any change(s) to the cell-culture process should be demonstrated not to affect product quality and should be discussed with NRA. In the manufacture of monoclonal antibodies cloning of hybridoma cultures is particularly important to ensure that a single product is generated since inclusion of more than one hybridoma cell type could lead to a mixture of different antibody specificities and classes being present.

The details of cloning and selection may vary, depending on the practices of individual manufacturers and should be discussed with the NRA/NCL.

For example, in the early stages of cell-line development, a number of different recombinant vector systems and cell lines may be used. This will essentially be a research activity, but the cell lines and vectors should originate from well characterized and qualified sources and the cells from an appropriately qualified seed stock or MCB, which will usually be 'in-house' host cells and vectors. The most promising cell/vector combination will then be used to generate a large number of clones (100s -1000s) after transfecting the culture with rDNA. Typically, these clones will be screened based on their productivity, and a number with the highest productivity (10-50) will be taken forward for further evaluation. Further testing will then be used to select a small number (1-5) for establishment as small pre-master cell banks, and a final selection will be made, often based on stability characteristics and amenability to scale-up, before finally generating a MCB and WCB. Throughout the process, only well characterized and traceable growth media and other critical reagents will be used (usually the same as for the MCB), and cryopreserved stocks of all working clones will be made at appropriate stages in the development process (Figure 1).

Figure 1. Simplified Schematic of an Example of the Development of a Genetically Modified Cell Line

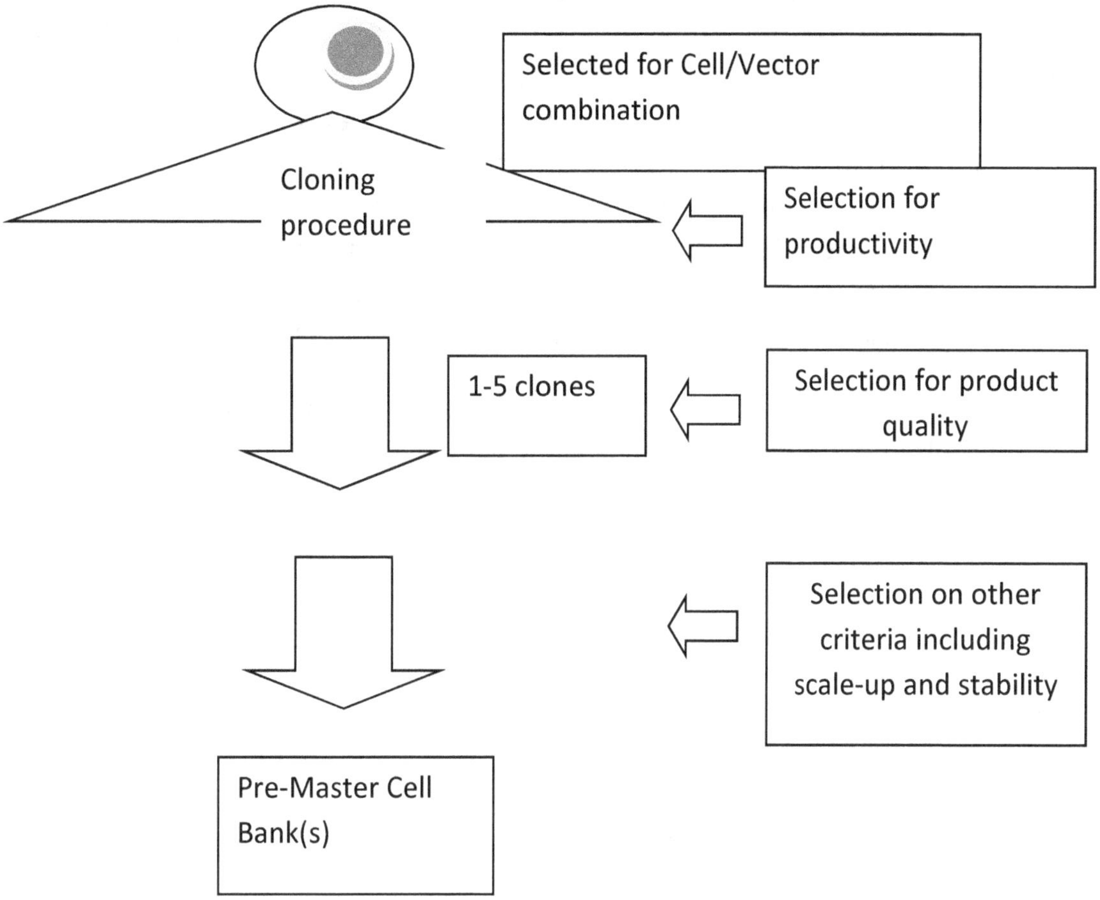

In the process of cloning a cell culture, single cells should be selected for expansion. The cloning procedure should be carefully documented, including the provenance of the original culture, the cloning protocol, and reagents used. Cloning by one round of limiting dilution will not necessarily guarantee derivation from single cells; additional subcloning steps should be performed. Alternatively or in addition to limiting dilution steps the cloning procedure can include more recent technology such as single cell sorting and arraying, or colony picking from dilute seeds into semi-solid media. In any case, the cloning procedure should be fully documented, accompanied by imaging techniques and/or appropriate statistics. For proteins derived from transfection with recombinant plasmid DNA technology a single, fully documented round of cloning is sufficient provided product homogeneity and consistent characteristics are demonstrated throughout the production process and within a defined cell age beyond the production process.

It is important to accurately document the establishment of each clone, which should also have a unique reference. Cryopreserved seed stocks of a significant number of clones should be established at an early stage. The clones can then be compared in parallel with the parental culture to establish candidate clones with the best overall characteristics for delivery of the desired product. The criteria used in the evaluation of the clone selected for production should include: genomic and phenotypic stability, growth rate, achievable product levels, and integrity/stability of the product. The evaluation of early candidate clones should generate sufficient information for the manufacturer to make an informed decision on the selection of the most promising clone(s) for further development. Where genetically engineered cell clones are under evaluation, these criteria also should include stability of integrated rDNA. The details of this process could vary depending on a number of factors including the nature of the host cell, desired characteristics of the product and the manufacturer's local procedures.

It is important to bear in mind that even following single cell cloning, epigenetic variation can result in a cloned culture showing evidence of heterogeneity (*i.e.*, more than one clone). This should not preclude the use of such a culture for production, unless there are indications of instability that could affect the quality and/or safety of the final product.

A.2.5 *Special considerations for neural cell types*

Agents causing transmissible spongiform encephalopathies have been propagated in certain cells. At the time of writing, the phenomenon has been observed only with very specific pairs of agents and cells; no cell line has been identified that will replicate all agents, although one has been described that seems to be infectable by many strains of scrapie. The phenomenon is unpredictable, except that if the line does not express the PrP protein, it may be assumed to be impossible to infect it, and experience to date suggests that infection is not commonly observed or easy to maintain. On the other hand, the cell types that can be infected include fibroblastic lines as well as neuronal cells. The cells have usually been of murine origin, because the infecting agents are usually mouse-adapted scrapie. The fact that certain cells can be infected with certain agents is proof of principle that cell lines may be infected so that exposure of cells to sources potentially contaminated with the agents is a concern. The scale of the risk is difficult to judge, and it is recommended that with respect to safety considerations and TSEs attention focuses on the selection and documentation of the cell culture reagents and other materials that come into intimate contact with the cells to provide assurance that they are not contaminated. Strategies to accomplish this are given in section B.11.4.

A.3 Selection of source materials

A.3.1 *Introduction*

All materials should be subjected to a risk assessment and testing when necessary, in particular, raw materials derived from humans and animals, which can be a primary source for the introduction of adventitious agents into the production of biologicals. Therefore, careful attention should be paid to their sourcing, production, handling, testing, and quality control. All cell-culture materials of biological origin that come into intimate contact with the cells during the establishment of cell cultures, derivation of a new cell line (if any), banking procedures (if any), and production, should be subjected to appropriate tests, as indicated by risk assessment, for quality and freedom from contamination by microbial agents to evaluate their acceptability for use in production. It is important to evaluate the microbiological risks represented by each individual human- and animal-derived reagent used in a cell culture production process, and it should address: (a) geographical origin; (b) species of origin; (c) general microbiological potential hazards including a consideration of the medical history for human-derived reagents; (d) husbandry/screening of donor animals; (e) testing performed on the product, including Certificates of Analysis (if any); and (f) the capacity for the preparation, purification, and sterilization procedures (if any) used to remove or inactivate contaminants [54]. Other reagents of biological, but of non-animal origin, may also present risks to product safety, and these are discussed further in section A.3.4

Recombinant protein technology now provides many materials formerly derived directly from animal or human sources. While this eliminates obvious virological risks from donors, the manufacturing process used for the recombinant proteins should be analyzed for any materials of biological origin and any associated hazards that may need to be addressed, as indicated in items a-f above.

The NRA/NCL should approve source(s) of animal-derived raw materials, such as serum and trypsin. These materials should comply with the guidelines on tissue infectivity distribution of TSEs [55]. They should be subjected to appropriate tests for quality and freedom from contamination by microbial agents to evaluate their acceptability for use in production. Their origin should be documented to ensure that the sources are from geographical regions with acceptable levels of microbiological risk (*e.g.*, freedom from foot and mouth disease virus or bovine spongiform encephalopathies). In addition, documentation should be gathered on their manufacturing history, production, quality control, and any final or supplementary processing that could affect quality and safety, such as blending and aliquoting of serum batches. In addition, controls should be in place to prevent cross-contamination of one material with another (*e.g.*, bovine material in a porcine product).

The reduction and elimination from the manufacturing process of raw materials derived from animals and humans is encouraged, where feasible.

For some human- and animal-derived raw materials used in the cell culture medium, such as insulin or transferrin, validation of the production process for the elimination of viruses can substitute for virus detection tests, when justified.

Animal-derived reagents, such as trypsin and serum, which would be substantially damaged or destroyed in physical sterilization processes, including heat and irradiation, present the most likely microbiological hazards to cell-culture processes. Batches of reagents, such as trypsin and bovine serum, have been known to contain Mycoplasma species and sometimes more than one viral contaminant. Certain contaminants also have been shown to infect cells in culture. The processing environment is also a common source of microbiological contamination and should be controlled to minimize this risk and prevent growth of contaminants.

A.3.2 *Serum and other bovine-derived materials used in cell culture media*

The source(s) of serum of bovine origin should be approved by the NRA/NCL. The responsibility for ensuring the quality of the serum used in the manufacture of cell banks and biologicals rests with the biologicals' manufacturer. This can be accomplished in more than one manner. The manufacturer might conduct adventitious agent testing and perform inactivation of the serum after purchase from the serum manufacturer. Alternatively, the manufacturer might qualify their serum vendor and only purchase serum from suppliers after conducting thorough and on-going audits of the serum supplier to ensure that they have properly performed the manufacture, quality control, and validation necessary to achieve the level of quality of the serum required for the biological it is being used to produce. In some cases, certificates of analysis may then be considered sufficient. Some combination of these approaches might be optimal, and the strategy taken should be considered in evaluating risk. Consultation with the NRA/NCL also might be advisable.

Serum and other bovine-derived materials should be tested for adventitious agents, such as bacteria, fungi, mycoplasmas, and viruses, prior to use in the production of MCBs and WCBs and in the manufacture of biologicals. Particular consideration should be given to those viruses that could be introduced from bovine-derived materials and that could be zoonotic or oncogenic (*e.g.*, bovine viral diarrhea virus, bovine polyoma virus, bovine circoviruses, rabies virus, bovine adenoviruses, bovine parvoviruses, bovine respiratory syncytial virus, infectious bovine rhinotracheitis virus, bovine parainfluenza virus type 3, reovirus 3, Cache Valley virus, bluetongue virus, and epizootic hemorrhagic disease virus). In addition, consideration should be given to risk-mitigation strategies, such as inactivation by heat or irradiation, to ensure that adventitious agents that were not detected in the manufacture and quality control of the serum would be inactivated to a degree acceptable to the NRA/NCL. If irradiation or other inactivation (*e.g.*, heat sterilization) methods are used in the manufacture of the serum, the tests for adventitious agents should be performed prior to inactivation to enhance the opportunity for detecting the contamination. If evidence of viral contamination is found in any of the tests, the serum is acceptable only if the virus is identified and shown to be present in an amount that has been shown in a validation study to be effectively inactivated. For serum that is not to be subjected to a virus inactivation / removal procedure, if evidence of viral contamination is found in any tests, generally, the serum would not be acceptable. If the manufacturer chooses to use serum that has not been inactivated, thorough testing of the serum for adventitious agents using current best practices should be undertaken. If any agents are identified, the cell banks made in this manner should be shown to be free of the identified virus(es).

If irradiation is used, it is important to ensure that a reproducible dose is delivered to all batches and the component units of each batch. The irradiation dose must be low enough so that the biological properties of the reagents are retained while being high enough to reduce virological risk. Therefore, irradiation delivered at such a dose may not be a sterilizing dose.

If serum was used in the establishment or passage history of the animal-cell substrate prior to banking by the manufacturer, the cell bank (MCB or WCB) and/or cells at or beyond the level of production should be tested for adventitious agents of the species (*e.g.*, bovine) of serum used in the establishment and passage history of the cell substrate. If serum is not used in the production of the subsequent stages, then this testing would not need to be repeated on those subsequent stages once the cell bank has been tested and considered free of bovine (or whichever species of serum that was used) adventitious agents.

Methods used to test for bovine viruses should be approved by the NRA/NCL. Details of the methods are provided in Appendix 1. Infectivity assays are used as the primary screening method and have resulted in the detection of bovine viral diarrhea virus, reovirus 3, Cache Valley virus, bluetongue virus, and epizootic hemorrhagic disease virus amongst others. However, it should be noted that, in general, the infectivity screening assay methods described here do not readily detect some of the viruses (*e.g.*, bovine polyomaviruses) that can be frequent contaminants of serum. Additional methods may need to be considered, such as the nucleic acid amplification technique (NAAT), although the presence of viral genomic sequences is not necessarily indicative of infectious virus. In those cases, specific infectivity assays designed to detect the virus of concern (*e.g.*, bovine polyomavirus) may need to be considered.

A second factor in screening serum is the limited sample volume used compared with the batch size, which may be on the order of 1000 liters, which comes from the pooling of serum from many animals. Consequently, infectious viruses can be missed in the serum lot testing, and consideration should be given to direct screening of the cell bank for bovine viruses. These assays could include, in addition to the general screening procedure, NAAT for the presence of bovine viruses that may infect the cell substrate but undergo abortive and/or transforming infections. Virus families of particular concern in this latter regard include polyomaviruses, herpesviruses, circoviruses, anelloviruses, and adenoviruses.

General screening assays for the detection of infectious viruses in serum or cell substrates involve the use of at least one indicator cell line, such as bovine turbinate cells, permissive for the replication of bovine viral diarrhea virus (BVDV). A second cell line such as Vero also should be employed to broaden the detection range. Before initiating screening it may be necessary to evaluate the serum for the presence of antibody, particularly to BVDV, that could mask the presence of infectious virus.

Typically, indicator cells are cultured in the presence of the serum under test for 21 to 28 days, passaging the cells as necessary. During this period, the cells are regularly examined for the presence of cytopathic effect (CPE) indicative of virus infection. At the end of the observation period, which should not be less than 7 days after the last sub-culture, the cells are stained to detect CPE that may have been missed during observation of the living cells. Additional endpoint assays should include haemadsorbtion and haemagglutination at both 4°C and a higher temperature such as 20 to 25°C and also immunofluorescence assays (IFA) for specific viruses. Appropriate controls should be used for each assay, like bovine parainfluenza virus type 3 for haemadsorbtion. IFA is particularly important for BVDV, as non-cytopathic BVDV may be present in the serum. IFA endpoints are also used to detect other viruses that may be determined by geographical considerations like adenoviruses, bovine parvovirus, bluetongue virus, bovine syncytial virus, reovirus type 3 and unlikely, but serious contaminants like rabies virus.

If serum from another species is used (*i.e.*, other than bovine), the NRA/NCL should be consulted regarding acceptable testing methods for that species.

A.3.3 *Trypsin and other porcine-derived materials used for preparing cell cultures*

Trypsin used for preparing cell cultures should be tested for cultivable bacteria, fungi, mycoplasmas and infectious viruses, including bovine or porcine parvoviruses, as appropriate. The methods used to ensure this should be approved by the NRA/NCL.

In some countries, irradiation is used to inactivate potential contaminant viruses. If irradiation is used, it is important to ensure that a reproducible dose is delivered to all batches and the component units of each batch. The irradiation dose must be low enough so that the biological properties of the reagents are retained while being high enough to reduce virological risk. Therefore, irradiation cannot be considered a sterilizing process.

The quality of the trypsin, like serum, is the responsibility of the biologicals' manufacturer (see section A.4.2). Recombinant trypsin is available and should be considered; however it should not be assumed to be free of risk of contamination and should be subject to the usual considerations for any reagent of biological origin (see A.4.1).

Like serum batches that are derived from many animals, trypsin batches are prepared from the pancreases of many animals. Most batches of porcine trypsin contain genetic sequences of porcine parvovirus 1 and porcine circoviruses and should be therefore treated to inactivate any virus potentially present in a manner accepted by the NRA/NCL. It is acknowledged however that these viruses are relatively resistant to inactivation [56]. If trypsin from another species is used, the NRA/NCL should be consulted regarding acceptable testing methods.

General screening assays for the detection of infectious porcine viruses in trypsin or cell substrates involve the use of at least one indicator cell line, such as porcine testes cells or Vero cells, permissive for the replication of porcine viruses. Typically, indicator cell cultures would be incubated for 14 days with a sub-culture onto fresh test cells for an additional 14 days. Specific endpoint detection methods like IFA or PCR may be required in addition to observation for cytopathic effect periodically throughout the culture period and more general endpoint detection methods like hemadsorption and/or hemagglutination.

Furthermore, if trypsin was used in the establishment or passage history of the animal-cell substrate prior to banking by the manufacturer, the cell bank (MCB or WCB) should be tested for porcine parvovirus, or appropriate adventitious viruses relevant to the species of origin of the trypsin used in the establishment and passage history of the cell substrate. If trypsin is not used in the production of the subsequent stages, then this testing would not need to be repeated on those subsequent stages once the cell bank has been tested and considered free of porcine parvovirus (or relevant agents).

Consideration should be given to screening for other agents such as porcine circoviruses. Molecular methods, such as PCR, may be used for such purposes.

Testing of cells exposed to trypsin or of other porcine-derived materials might entail testing for more than porcine parvovirus or porcine circoviruses. For example, testing for porcine adenovirus, transmissible gastroenteritis virus, porcine haemagglutinating encephalitis virus, bovine viral diarrhea virus, reoviruses, rabies virus, porcine anellovirus, porcine hokovirus, porcine bocavirus, porcine hepatitis E virus, porcine reproductive and respiratory syndrome virus, encephalomyocarditis virus, and potentially other viruses might be appropriate. Particular consideration should be given to those viruses that could be introduced from the porcine-derived material and that could be zoonotic or oncogenic. Additionally, tests for bacterial and fungal sterility and mycoplasmas depending on the type of porcine-derived material should be conducted. The NRA/NCL should be consulted in this regard.

A.3.4 *Medium supplements and general cell culture reagents derived from other sources used for preparing cell cultures*

Medium supplements derived from other species should be quality controlled from the perspective of adventitious agents. Consideration should be given to whether recombinant-derived medium supplements were exposed to animal-derived materials during their manufacture, and if so, evaluated for the potential to introduce adventitious agents into the manufacture of the cell banks and biological products. Testing for adventitious agents should assess viruses relevant to the species from which the supplement was derived. The NRA/NCL should be consulted in this regard.

Generally, medium supplements should not be obtained from human source materials. In particular, human serum should not be used. However, in special circumstances, and in agreement with the NRA/NCL, the use of human-derived supplements may be permitted. If human serum albumin is used at any stage of product manufacture, the NRA/NCL should be consulted regarding the requirements, as these may differ from country to country. However, as a minimum, it should meet the revised Requirements for Biological Substances No. 27 [56], as well as the guidelines on tissue infectivity distribution of TSEs [55].

Recombinant human albumin is commercially available and should be considered; however, it should not be assumed to be free of risk of contamination and should be subject to the usual considerations for any reagent of biological origin (see A.3).

As for other cell-culture reagents, it is important to establish traceability and assess and reduce microbiological risks as described in section A.3.

A variety of cell-culture reagents of biological origin are available that are derived from non-animal sources, including a range of aquatic organisms, plants and algae. In such cases, the exact hazards involved may be uncertain and unfamiliar. The microbiological risks may be substantially different to those involved in animal-derived reagents (see section A.3.1-A.3.3), and other hazards may arise, such as immunogenic, mitogenic and allergenic properties of the reagent and its components. For example, plant-derived material may carry an increased risk of mycoplasma and mycobacteria contamination.

A.4 Certification of cell banks by the manufacturer

It is vital that the manufacturer has secured a body of information on the cell substrate that demonstrates clearly the origin or provenance of the culture and how the cell banks intended for production (MCB and WCB) were established, characterized, and tested. This should provide all the information required to demonstrate the suitability of that cell substrate and the established cell banks for the manufacture of biological products.

A.4.1 *Cell line data*

Each new cell line should have an associated body of data, which will increase as the cell line is established and developed for manufacturing purposes. This data set is vital to demonstrate the suitability for use of the cells and should provide information on cell provenance (donor information and any relevant details on ethical procurement), cell line derivation, culture history, culture conditions (including reagents), early stage safety evaluation data, banking and cell bank characterization, and safety testing. This information should be available to the NRA/NCL for approval of the cells used in manufacture.

A.4.2 *Certification by the manufacturer of primary cell cultures*

Full traceability should be established for PCCs to the animals of origin, husbandry conditions, veterinary inspection, vaccinations (if any), procedures for administering anesthesia and cell harvesting, the reagents and procedures used in the preparation of primary cultures, and the environmental conditions under which they were prepared. Extensive testing should be performed, and this should be documented.

It is important to define the batch or lot of cells used in each individual manufacturing process. For production purposes, a batch or lot is a culture of primary cells derived from single or multiple animals that has been subjected to a common process of tissue retrieval, disaggregation, and processing leading to a single culture preparation of cells. Lots may be prepared by harvesting and pooling cells in different ways, but the cell-processing procedures should be reproducible, and it is especially important to monitor cultures carefully for evidence of adverse change in the cell culture and microbiological contamination. Prior to any culture pooling, cells should be examined for acceptability for production. Acceptability criteria should be established and should include testing for microbiological contamination and the general condition of the cells (*e.g.*, morphology, number, and viability of the cells). Failure to detect and eliminate atypical (*i.e.*, potentially virally infected) or grossly contaminated cells will put the entire production run at risk and could compromise the safety of the product. Cells showing an unacceptably high proportion of dead or atypical cells should not be used, and ideally microbiological testing should be completed and passed before the cells are used.

The preparation of cell lots for manufacture should be carefully documented to provide full traceability from the animal donor(s) to production. Any pooling of cells should be clearly recorded, as should any deviation from standard operating procedures, as required by GMPs. In addition, any observations of variation between batches should be recorded, even where such observations would not necessarily lead to rejection of those batches. Such information may prove valuable in ongoing optimization and improvement of the production process.

A.4.3 *Certification by the manufacturer of diploid, continuous, and stem cell lines*

All cell lines used for biologicals production should have data available, as indicated in A.4.1.

The original PDL (or passage number, if the PDL is unknown) of the cell seed should be recorded.

For cell lines of human origin, if possible the medical history of the individual from whom the cell line was derived should be evaluated in order to better assess potential risks and the suitability of the cell line.

For SCLs, morphology continues to be an important characteristic, and representative images and immune-phenotypic profiles of undifferentiated and differentiated cells should be available for comparison.

A.5 Cryopreservation and Cell Banking

A.5.1 *Cryopreservation*

When cells are banked, the successful preservation of cells at ultra-low temperatures is critical to the efficient delivery of good quality cultures (*i.e.*, high viability cultures of the required characteristics). The need to prepare large stocks of frozen vials of cells for cell banks is especially challenging, and a number of key principles should be adopted:

- A method that meets current best practice for cell-culture preservation should be used (for example, see ref 58)

- The cooling profile achieved for the cells being frozen should be defined, and the same cooling process should be used for each separate preservation process (*i.e.*, an SOP should include documentation of the cooling process in the batch record.)
- Each preservation process should be recorded
- As a general guide, only cell cultures that are predominantly in the exponential phase of growth should be used. Cells in such cultures tend to have a low ratio of cytoplasm to nucleus (v/v) and should be more amenable to successful cryopreservation. It is unwise to use cells predominantly in the 'lag' phase very early after passage or in the 'stationary' phase when the culture has reached its highest density of cells
- For each bank, cells pooled from a single expanded culture (*i.e.*, not from a range of cultures established at different times post seeding or different PDLs) should be used and mixed prior to aliquotting to ensure homogeneity
- The number of cells per vial should be adequate to recover a representative culture (*e.g.*, 5-10 x 10^6 in a 1-mL aliquot)
- For new cell banks, antimicrobials should not be used in cell cultures to be banked, except where this can be justified for early PDL cultures which may carry contamination from tissue harvesting or recombinant cells which require antibiotic selection, and when necessary for the genetic stability of recombinant cell lines. In any case, if antimicrobials are used, they should not be penicillin or any other beta-lactam drug
- When a stock of cells has been frozen, a sample should be recovered to confirm it has retained viability and the results recorded. It is also important to establish the degree of homogeneity within the cell bank. Recovery of a sufficient percentage (*e.g.*, 1% or as recommended by the NRA) of vials representative of the beginning, middle and end of the cryopreservation process should be demonstrated to give confidence in the production process based on the use of that cell bank. Ultimately, stability (see B.3) and integrity of cryopreserved vials is demonstrated when the vials are thawed from production and demonstrated to produce the intended product at scale (see also B.7).
- Cell bank cryostorage vessels should be monitored and maintained to enable demonstration of a highly stable storage environment for cell banks. Access to such vessels may cause temperature cycling, which in extreme cases can cause loss of viability. It is therefore prudent to establish a stability-testing programme involving recovery of cells periodically where the frequency of recovery relates to risk of temperature cycling. New developments in remote monitoring of individual vials may in future eliminate the need for stability testing.

A.5.2 *Cell Banking*

When DCLs, SCLs, or CCLs are used for production of a biological, a cell-bank system should be used. Such banks should be approved by and registered with the NRA/NCL as part of the product approval process. The source of cells used in cell banking and production is a critical factor in biological product development and manufacture. It is highly desirable to obtain cells from sources with a documented history and traceability to the originator of the cell line.

After a sample of the original seed stock is obtained, an early stage pre-master bank of just a few vials should be established. One or more vials of that pre-master bank are used to establish the MCB. The WCB is derived by expansion of one or more containers of the MCB. The WCB should be qualified for yielding cell cultures that are acceptable for use in manufacturing a biological product.

When using early PDLs from primary cultures for production processes, the preparation of a cell bank should be considered on a case-by-case basis. This approach has significant benefits, as it gives great flexibility in the timing of the production process, permits quality control and safety testing to be completed prior to use, and reduces the overall burden of testing required in the production process.

Cell banks should be characterized as specified in Part B of this guidance and according to any other currently applicable and future guidance published by WHO. The testing performed on a replacement MCB (derived from the same cell clone or from an existing MCB or WCB) is the same as for the initial MCB unless a justified exception is made. Efforts to detect contaminating viruses and other microbial agents constitute a key element in the characterization of cell banks.

Having been cryopreserved by qualified methods (see A.5.1), both the MCB and WCB should be stored frozen under defined conditions, such as in the vapour or liquid phase of liquid nitrogen. The location, identity, and inventory of individual cryovials or ampoules of cells should be thoroughly documented. It is recommended that the MCB and WCB each be stored in at least two widely separated areas within the production facility and/or in geographically distinct locations to assure continued ability to manufacture product in the event of a facility catastrophe. When cryopreserved cells are transferred to a remote site, it is important to use qualified shipping containers and to monitor transfers with probes to detect temperature excursions. All containers are treated identically and, once removed from storage, usually are not returned to the stock. The second storage site should operate under an equivalent standard of quality assurance as the primary site.

A.5.3 *WHO reference cell banks (RCBs)*

The principle of establishing RCBs under WHO auspices is one that offers potential solutions to future challenges for the development of vaccines and biotherapeutics in developing regions. However, WHO does not intend that cells supplied to manufacturers from any RCB be used as a MCB. The purpose of WHO RCBs is to provide well characterized cell seed material for the generation of a MCB by manufacturers with the expectation that such MCBs will comply with this guidance document and be fully characterized.

The WHO RCBs provide key advantages for vaccine development worldwide that include:

- Traceability to origin of cells and derivation of cell line and materials used in preparation of seed stock
- Subjected to open international scientific scrutiny and collaborative technical investigations into the characteristics of the cells and the presence of adventitious agents
- Results of characterization were peer reviewed and published
- Investigations evaluated under auspices of WHO expert review and qualified as suitable for use in vaccine production
- Supply of cells free of any adverse Intellectual Property 'reach through' on final products
- Single source of cells with growing and scientifically and technically updated body of safety-testing data and safe history of use, giving increasing confidence for manufacturers, regulators, and public policy makers

The Vero cell line is the most widely used continuous cell line for the production of viral vaccines over the last two decades. The WHO Vero RCB 10-87 was established in 1987 and was subjected to a broad range of tests to establish its suitability for vaccine production. This WHO RCB provides a unique resource for the development of future biological medicines where a cell substrate with a safe and reliable history of use is desired. A comprehensive review of the characterization of the WHO Vero 10-87 seed lot was conducted recently, and a detailed overview is provided on the WHO website at (http://www.who.int/biologicals/).

As concluded by an expert review in 2002, the WHO Vero RCB 10-87 is not considered suitable for direct use as MCB material. However, the WHO Vero RCB 10-87 is considered suitable for use as a cell seed for generating a MCB, and its status has changed from "WHO Vero cell bank 10-87" to "WHO Vero reference cell bank 10-87" .

The WHO Vero RCB 10-87 is stored in the European Collection of Animal Cell Cultures (ECACC, www.hpacultures.org.uk), Salisbury, Wiltshire, UK, and the American Type Culture Collection (ATCC, www.atcc.org), in Manassas, Virginia, USA. These public service culture collections have distributed ampoules under agreements with the WHO to numerous manufacturers and other users. The WHO Vero RCB 10-87 is the property of WHO, and there are no constraints relating to intellectual property rights. The WHO Vero RCB 10-87 is available free of charge on application to WHO. However, due to the limited number of vials remaining, distribution of these vials is restricted solely for the production of vaccines and other biologicals. Potential replacement of the WHO Vero RCB 10-87 is currently under consideration.

WHO also has overseen the establishment of seed stocks of MRC-5 for the production of vaccines. The WHO MRC-5 RCB was established in 2007 because of stability issues associated with the original vials of MRC-5 cells, which dated to 1966. This RCB was prepared in a qualified cleanroom environment and subjected to specified quality-control testing endorsed by the ECBS.

Part B. Recommendations for the characterization of cell banks of animal cell substrates

B.1 General considerations

Since the 1986 Study Group report, advances in science and technology have led to an expanded range of animal cell types that are used for the production of biological products. In some cases, these new cell types provide significantly higher yields of product at less cost, while in other cases they provide the only means by which a commercially viable product can be manufactured. Many products manufactured in CCLs of various types have been approved, and some examples are listed in Table 1.

Table 1. Examples of Approved Biological Products Derived from CCLs

Product Class	Product (disease)	Cell Line
Therapeutic	Factor VIII (hemophilia)	CHO
	Factor VIIa (hemophilia)	BHK-21
	Monoclonal antibody (various disease)	CHO and murine myeloma (NS0 and SP2/0)
Prophylactic	Poliovirus vaccine	Vero
	Rotavirus vaccine	Vero
	Rabies vaccine	Vero
	JE vaccine	Vero
	Human papillomavirus vaccine	Sf-9
	Influenza vaccine	MDCK

Many more products are currently in development, and some use highly tumourigenic cells (*e.g.*, HeLa; some banks of MDCK), and some involve sources previously unused in production such as insect cells. Examples are listed in Table 2.

Table 2. Examples of Biological Products in Development Derived from CCLs

Product Class	Product (disease)	Cell Line(s)
Therapeutic	> 50% of products in development use CHO or murine myeloma cells as the cell substrate [59]. Monoclonal antibodies are generally produced using CHO, SP2/0, PER.C6, and NS0 cells [60].	
Prophylactic	HIV vaccines	CHO, Vero, PER.C6, 293ORF6, HER96, HeLa
	Herpes Simplex Type 2 vaccine	CHO
	Influenza vaccines	SF9, Vero, PER.C6
	Rabies vaccine	S2

CCLs may have biochemical, biological, and genetic characteristics that differ from PCCs or DCLs and that may impose a risk for the recipients of biologicals derived from them. In particular, they may produce transforming proteins, and may contain potentially oncogenic DNA and viral genes. In some cases, CCLs may cause tumours when inoculated into animals. Non-tumourigenic cells (*e.g.*, PCCs and DCLs) had been thought to be intrinsically safer than tumourigenic cells. Where tumourigenic cells have been used in the past (*e.g.*, CHO for recombinant proteins), high degrees of purity have been required with a special emphasis on reduction in quantity of DNA. When not possible to completely remove the amount of DNA to below the detection limit, emphasis has been on a reduction in size or other approaches to inactivate rcDNA and rcRNA (*e.g.*, beta-propiolactone (BPL) for rabies vaccine).

Manufacturers considering the use of CCLs should be aware of the need to develop, evaluate, and validate efficient methods for purification as an essential element of any product-development programme. However, a minimally purified product, such as certain viral vaccines (*e.g.*, polio), may be acceptable when produced in a CCL such as Vero when data are developed to support the safety of the product. Such data would include extensive characterization of the MCB or the WCB and of the product itself.

While tumourigenicity tests have been part of the characterization of CCLs, they comprise only one element in an array of tests, the results of which must be taken into account when assessing the safety of a biological produced in a given cell substrate. For example, if a CCL is positive in a tumourigenicity test, and if the CCL is to be used for the production of a live viral vaccine, an evaluation of the oncogenic potential of the cells might be requested by the NRA/ NCL to characterize the cellular DNA and to detect oncogenic viruses that might be present. Such studies should be discussed with the NRA/NCL.

Evidence should be provided for any animal-cell line proposed for use as a substrate for the manufacture of a biological product demonstrating that it is free from cultivable bacteria, mycoplasmas, fungi, and infectious viruses, including potentially oncogenic agents to the limits of the assay's detection capabilities. Special attention should be given to viruses that commonly contaminate the animal species from which the cell line is derived, and to cell-culture reagents of biological origin. The cell seed should preferably be free from all microbial agents. However, certain CCLs may express endogenous retroviruses. Tests capable of detecting such agents should be carried out on cells grown under cell-culture conditions that mimic those used during production, and the levels of viral particles should be quantified. Viral contaminants in a MCB and WCB should be shown to be inactivated and/or removed by the purification procedure used in production. The validation of the purification procedure used is considered essential.

The characterization of any DCL, SCL or CCL to be used for the production of biologicals should include: a) a history of the cell line (*i.e.*, provenance) and a detailed description of the production of the cell banks, including methods and reagents used during culture, PDL, storage conditions, viability after thawing, and growth characteristics; b) the results of tests for infectious agents; c) distinguishing features of the cells, such as biochemical, immunological, genetic, or cytogenetic patterns, that allow them to be clearly distinguished from other cell lines; and d) the results of tests for tumourigenicity, including data from the scientific literature. Additional consideration should be given to products derived from cells that contain known viral sequences (*e.g.*, Namalwa, HeLa, 293, and PER.C6).

The recommendations that follow are intended as guidance for NRAs, NCLs, and manufacturers as the minimum amount of data on the cell substrate that should be available when considering a new biological product for approval. The amount of data that may be required at various stages of clinical development of the product should be discussed and agreed with the NRA/NCL at each step of the program.

B.2 Identity

Cell Banks should be authenticated by a cell-identification method approved by the NRA/NCL. Wherever practicable, methods for identity testing should be used that give specific identification of the cell line to confirm that no switching or major cross-contamination of cultures has arisen during cell banking and production. A number of the commonly used identity testing methods are compared in Table 3. In the case of human cells, genetic tests such as DNA profiling (*e.g.*, Short Tandem Repeat analysis, multiple Single Nucleotide Polymorphisms) will give a profile that is at least specific to the individual from whom the cells were isolated. Another test that might be used for human cells is HLA typing. Other tests that may be used but tend to be less specific include isoenzyme analysis and karyology, which may be particularly useful where there are characteristic marker chromosomes. However, where more specific genetic markers are available, they should be considered. It is not unexpected that a small proportion of cell lines, particularly those which are transformed, may show alterations to the expected identity profile. This has been observed in isoenzyme analysis where in rare cases a particular cell line may show a consistently different profile to that expected for the species of origin and is also a general issue relating to the effect of genetic instability for molecular identity testing techniques. Such effects in standard technologies are rare and may also arise with the implementation of new techniques. The implications of any unexpected results should be discussed with the NRA/NCL. For recombinant-protein products, cell line identity testing should also include tests for vector integrity, expression plasmid copy number, insertions, deletions, number of integration sites, the percentage of host cells retaining the expression system, verification of protein-coding sequences, and protein-production levels.

Table 3. Identity Testing for Mammalian Cell Lines

Technique	Advantages	Disadvantages
Karyology (especially useful for DCLs and SCLs)	Gives whole chromosomal genome visualisation and analysis that can identify species of origin for a very wide range of species using the same methodology. Newer methods include, Spectral Karyotyping which involves the use of probes, labeled with fluorescent dyes. The probes paint the chromosomes yielding different colors in specific areas. Spectral karyotyping is able to detect translocations not recognizable by traditional banding methods.	Results are generally not specific to the individual of origin (*i.e.*, usually species specificity) although certain cell lines may have marker chromosomes that are readily recognised. Giemsa banding requires special expertise and is labour intensive. Standard analysis of 10-20 metaphase spreads is insensitive for detecting contaminating cells.
Isoenzyme analysis	Determination of species of origin within a few hours	Analysis for 4-6 isoenzyme activities will generally identify species of origin, but is not specific to the individual of origin
DNA profiling using variable number of tandem repeats (VNTR) analysis or other PCR method such as Exon Primed Intron Crossing-PCR (EPIC-PCR), or other techniques, such as restriction fragment length polymorphism (RFLP)	Short tandem repeats (STR) analysis by PCR is rapid and gives identity specific to the individual of origin. Commercial kits are available for a range of human populations. EPIC-PCR method is rapid and gives identity specific to the individual of origin. It provides the advantage of covering a broad spectrum of organisms and cell lines other than human cells	Some limited, but undefined, cross-reaction of human STR primers for primate cells

B.2.1 *Applicability*

Cell Banks: MCB and each WCB

Cell Types: DCL, SCL, CCL

B.3 Stability

The stability of cell banks during cryostorage and the genetic stability of cell lines and recombinant expression systems are key elements in a successful cell bank program.

B.3.1 *Stability during cryostorage*

Data should be generated to support the stability or suitability of the cell substrate and any recombinant expression system or necessary cell phenotype during cultivation to or beyond the limit of production, and to support the stability of the cryopreserved cell banks during storage. The latter may be demonstrated by successful manufacture of WCBs or production lots. Periodic testing for viability is not necessary if continuous-monitoring records for storage show no deviations out of specification, and periodic production runs are successful. If banks are used less than once every 5 years, then it may be prudent to generate data confirming suitability for manufacturing on a schedule that takes into account the storage condition once every 5 years.

B.3.1.1 *Applicability*

Cell Banks: MCB and WCB

Cell Types: DCL, SCL, CCL

B.3.2 *Genetic Stability*

Any form of genetic instability could potentially affect the quality of the final product and it will be important to know if the cells in culture are changing in a way that could affect the nature or safety of the product. Any features of the cell lines that might affect quality should be discussed with the NRA/NCL to ensure that tests used by the manufacturer to monitor genetic stability are adequate. The specific tests will vary according to the nature of the product, but the aim is to show consistency in the amount and characteristics of the product derived from cells within a few passages of the MCB or WCB with those derived from an ECB or EOPC. For recombinant protein products, emphasis will be on the protein sequence and post translational modifications.

For cell lines containing DNA expression constructs, the stability of these constructs between the MCB/WCB and an ECB or EOPC should be determined. The copy number of the construct and, if relevant, the sites of chromosomal insertion should be determined. The latter is accomplished by sequencing into the cellular flanking regions, but methods like fluorescent *in situ* hybridization may provide useful additional information, particularly where concatamers of the gene insert are present at individual chromosomal loci. The sequence of the construct within the cells should be determined. With conventional sequencing, a consensus sequence is obtained, but with massively parallel sequencing, it is possible to determine the sequence of individual gene inserts or their transcripts.

Where proteins are derived from non-genetically modified cells, consistency in the yield and properties of the protein should be evaluated together with the sequence of the mRNA encoding the protein of interest.

Additional characterization of the cell-biological processes and responses during cultivation (for instance using global or targeted gene expression, proteomic or metabolic profiles and other phenotypic markers) might be useful in further developing a broad understanding of the cell substrate.

Appropriate methods should be applied to assure that cell age is correctly assessed in the event that cell viability falls dramatically at any given step. Losses in viability are reflected in increased cultivation times to reach defined levels of growth.

The stability of cell function in terms of productivity within the production process also may need to be evaluated. Other stability studies may be performed where bioreactor methods are employed, especially where extended culture periods are involved.

B.3.2.1 *Applicability*

Cell Banks: MCB taken to EOPC/ECB

Cell Types: DCL, SCL, CCL

B.4 Sterility

(see B.11.3.1)

B.5 Viability

High level of viability of cryopreserved cells is important for efficient and reliable production. Typically, thawed cells should have viability levels in excess of 80%, though this is not always achieved and may depend on the cell line. Lower viabilities may still result in suitable growth recovery and in acceptable product qualities. In such cases, the data should be discussed with the NRA/NCL. A range of viability tests are available that measure different attributes of cell function (membrane integrity, metabolic activity, DNA replication). Under certain circumstances, such as pre-apoptotic cells excluding trypan blue, viability assays may give misleading results, and it is important to be aware of the exact information that a particular viability assay provides. Therefore, it is important to evaluate growth recovery of cryopreserved cells upon thawing.

For certain cell cultures such as hybridomas, where a membrane-integrity test is used, additional cell markers such as indicators of apoptosis should be studied in order to avoid significantly overestimating viability.

A suitable viability test should be selected for the cell substrate in question and typical test values established for cultures considered to be acceptable (see B.6, B.7). It may also be necessary to select alternative viability assays that are better suited to providing in-process viability data required during production, *e.g.*, lactate dehydrogenase levels in bioreactor systems.

B.5.1 *Applicability*

Cell Banks: MCB and WCB

Cell Types: DCL, SCL, CCL

B.6 Growth Characteristics

For the development of production processes, the growth characteristics of the production-cell line should be well understood to ensure consistency of production. Changes in these characteristics could indicate any one of a range of events. Accordingly, data on growth characteristics, such as viability, morphology, cell-doubling times, cloning and/or plating efficiency, if applicable, should be developed. For certain cell substrates, it may be appropriate to apply such tests in homogeneity testing (see B.7). Experiments to demonstrate homogeneity and growth characteristics may be combined, although the analysis should be carried out separately.

B.6.1 *Applicability*

Cell Banks: MCB and WCB

Cell Types: DCL, SCL, CCL

B.7 Homogeneity Testing

Each cell culture derived from a container of the WCB should perform in the same way (*i.e.*, within acceptable limits and provide the same number of viable cells of the same growth characteristics). In order to assure this, it is important to recover a proportion of containers from each cell bank and check their characteristics, as indicated in B.6. The number of containers tested should be discussed with the NRA/NCL and be broadly in line with those normally sampled to establish product consistency. Recovery of a sufficient percentage of vials representative of the beginning, middle and end of the aliquotting process should be demonstrated to give confidence in the production process based on the use of that cell bank. Ultimately, stability and integrity of cryopreserved vials are demonstrated when the vials are thawed for production and demonstrated to produce the intended product at scale. Instead of testing a portion of containers at different stages of the banking process, an alternative strategy to ensure the homogeneity of the banks can be used based on the validation of the process method for filling and freezing. Assessment of growth characteristics (B.6) and homogeneity testing are commonly combined experimentally; however, the analysis and interpretation of each should be distinct. It may be appropriate to also test homogeneity of the MCB to assure future WCBs are consistent with the first WCB.

B.7.1 *Applicability*

Cell Banks: MCB, WCB

Cell Types: DCL, SCL, CCL

B.8 Tumourigenicity

B.8.1 *General considerations*

Several *in vitro* test systems, such as cell growth in soft agar [61] and muscle organ culture [62], have been explored as alternatives to *in vivo* tests for tumourigenicity; however, correlations with *in vivo* tests have been imperfect or the alternative tests have been technically difficult to perform. Therefore, *in vivo* tests remain the standard for assessing tumourigenicity.

Although WHO Requirements [1] have described acceptable approaches to tumourigenicity testing, a number of important aspects of such testing were not addressed. Therefore, a model protocol has been developed and is appended to this document. The major points included in the model protocol are listed below along with comments on each specific item.

A new diploid cell line (*i.e.*, other than WI-38, MRC-5, and FRhL-2) should be tested for tumourigenicity as part of the characterization of the cell line, but should not be required on a routine basis.

The tumourigenicity tests currently available are in mammalian species whose body temperatures and other physiologic factors are different from those of avian and insect species. Therefore, when the test is performed on avian or insect cells, the validity of the data is open to question unless a tumourigenic cell line of the species being tested is included as a positive control. The NRA/NCL may accept the results of an *in vitro* test such as growth in soft agar as a substitute for the *in vivo* test for avian and insect cell lines. However, as mentioned above, correlations of *in vitro* tests with *in vivo* tests are imperfect. This should be discussed with the NRA/NCL.

Many CCLs (*e.g.*, BHK-21, CHO, HeLa) are classified as tumourigenic because they possess the capacity to form tumors in immunosuppressed animals such as rodents.

Some CCLs become tumourigenic at high PDLs (*e.g.*, Vero), even though they do not possess this capacity at lower PDLs at which vaccine manufacture occurs. A critical feature regarding the pluripotency of embryonic SCLs, even though they display a diploid karyotype, is that they form tumours in immunocompromised mice.

The expression of a tumourigenic phenotype can be quite variable from one CCL to another, and even within different sub-lines of the same CCL. This range of variability, from non-tumourigenic, to weakly tumourigenic, to highly tumourigenic, has been viewed by some as indicating different degrees of risk when they are used as substrates for the manufacture of biological products [10,11].

If the CCL has already been demonstrated to be tumourigenic (*e.g.*, BHK-21, CHO, HEK293, Cl27), or if the class of cells to which it belongs is tumourigenic (*e.g.*, hybridomas, SCLs), it may not be necessary to perform additional tumourigenicity tests on cells used for the manufacture of therapeutic products. Such cell lines may be used as cell substrates for the production of biological products if the NRA/NCL has determined, based on characterization data as well as manufacturing data, that issues of purity, safety, and consistency have been addressed. A new cell line (DCL, SCL, or CCL) should be presumed to be tumourigenic unless data demonstrate that it is not. If a manufacturer proposes to characterize the cell line as non-tumourigenic, the following tests should be undertaken.

Cells from the MCB or WCB propagated to the proposed *in vitro* cell age used for production or beyond should be examined for tumourigenicity in a test approved by the NRA/NCL. The test should involve a comparison between the cell line and a suitable positive reference preparation (*e.g.*, HeLa cells) and a standardized procedure for evaluating results.

B.8.2 *Type of test animals*

A variety of animal systems have been used to assess the tumourigenic potential of cell lines. Table 4 lists several examples of such tests along with advantages and disadvantages of each. Because assessing the tumourigenic phenotype of a cell substrate requires the inoculation of xenogeneic or allogeneic cells, the test animal should be rendered deficient in cytotoxic T-lymphocyte (CTL) activity. This can be accomplished either by the use of rodents that are genetically immunocompromised (*e.g.*, nude mice, severe combined immunodeficiency (SCID) mice) or by inactivating the T-cell function with anti-thymocyte globulin (ATG), anti-thymocyte serum (ATS), or anti-lymphocyte serum (ALS). The use of animals with additional defects in NK-cell function, such as, the SCID-NOD mouse, the SCID-NOD-gamma mouse, and the CD3 epsilon mouse, has not yet been explored for cell-substrate evaluation, but they might offer some advantages. In addition to these systems, several other *in vivo* systems, such as the hamster cheek pouch model and ATG-treated non-human primates (NHPs), have been used in the past, but rarely at present.

Table 4. *In vivo* tests to assess the tumourigenic potential of inoculated cells

Test & Brief Description	Advantage(s)	Disadvantage(s)	Refs
Adult Athymic mouse (Nu/Nu genotype): Animals inoculated by the i.m. or s.c. route with cells to be tested	• Animals readily available • No immunosuppression required	• Higher frequency of spontaneous tumors than in other animal models that are not genetically immunosuppressed • Low sensitivity for assessing the metastatic potential of the inoculated cells	65
Newborn athymic mouse Animals inoculated by the s.c. route with cells to be tested	• No immunosuppression required • More sensitive than adults	• Low sensitivity for assessing the metastatic potential of the inoculated cells • Since litters include heterozygous mice, twice the number of animals must be inoculated in order to be sure that a sufficient number of homozygous mice have been included. • Cannibalism of newborns by the mother	66
SCID mouse: Animals receive subcutaneous, intra-dermal. or intra-kidney capsule inoculation of test cells	• No immunosuppression required • Potentially increased sensitivity • Animals readily available	• Highly susceptible to viral, bacterial, and fungal infections • Infections can affect the results and reproducibility of studies • Spontaneous thymic lymphomas may occur	64, 67
Newborn rat: Animals immunosuppressed with ATG followed by i.m. or s.c. inoculation of cells to be tested	• Animals readily available • Sensitive model for detecting metastases • Very low frequency of spontaneous tumor formation	• Standardized ATG not available as a commercial product • Careful qualification and characterization of the ATG is required to find the balance between immunosuppressive capacity and toxicity	65, 68
Newborn hamster or mouse: Animals immunosuppressed with ATS followed by i.m. or s.c. inoculation of cells to be tested	• Animals readily available	• Cannibalism of newborns by mother • Standardized ATS not available and difficult to balance toxicity vs. immunosuppressive capacity	69
Newborn hamster or mouse: Animals immunosuppressed with ALS followed by i.m. or s.c. inoculation of cells to be tested	• Increased sensitivity compared to the HCP test • Animals readily available	• Cannibalism of newborns by mother • Standardized ALS not available and difficult to balance toxicity vs. immunosuppressive capacity	70
Hamster cheek pouch (HCP): Animals immunosuppressed with cortisone followed by inoculation of the cells to be tested into the cheek pouch	• Animals readily available	• Lower sensitivity than newer models	71
Nonhuman primates: Animals immunosuppressed with ATG followed by inoculation of cells to be tested into the muscle of the four limbs	• Species closer to human	• Standardized ATG not available • Animals not readily available • Expense and limited availability preclude using large numbers • Animal welfare principles mandate against use of NHP if same results can be obtained from lower specie	72

Although all of the animal models listed in Table 4 have been used to assess the tumourigenicity of cells, several sensitivity parameters from studies using positive-control cells should be considered when attempting to compare the various *in vivo* tumourigenicity models: i) frequency of tumour formation; ii) time to appearance of tumors; iii) size of tumours; iv) lowest number of inoculated tumour cells that result in tumour formation; and v) metastatic tumor formation. Factors i, ii, iii, and v usually depend on the number of cells inoculated (*i.e.*, they are dose-dependent). In addition, the rate of spontaneous tumour formation should be considered. Although comparisons of two or more assays have been reported in the literature [63,69,73] none of them takes into account all of these factors, nor do they use the same tumourigenic cell lines. Thus, it is not possible to draw definitive conclusions on the relative sensitivity of the various tumourigenicity assays. Nevertheless, the following points appear to be generally accepted: i) the ATS-treated newborn rat and the ATG-treated nonhuman primate systems are the most sensitive to assess the metastatic potential of inoculated cells; ii) ATS and ATG provide better immunosuppression than ALS; iii) the nude mouse has a more well-defined level of immunosuppression than those that depend on ALS, ATS, or ATG, and inter-laboratory comparisons of nude mouse data are more likely to yield valid conclusions.

The overall experience during the past 30 years, and taking into consideration the points mentioned above, has led to the conclusion that the athymic nude mouse is an appropriate test system for determining the tumourigenic potential of cells proposed for use in the production of biologicals. The major advantages of the athymic nude mouse system are that it is easier to establish and standardize and is generally available, while the newborn rat system is more sensitive for assessing the metastatic potential of tumourigenic cells. In some cases, it might be preferable to use newborn athymic nude mice, as these animals are more sensitive than adults for the detection of weakly tumourigenic cells [64]. A tumourigenicity testing protocol using athymic nude mice is provided as Appendix 2. The animal system selected should be approved by the NRA/NCL.

B.8.3 *Point in the life history of the cells at which they should be tested*

Investigation of tumourigenicity should form part of the early evaluation of a new cell substrate for use in production. Cells from the MCB or WCB, propagated to the proposed *in vitro* cell age used for production or beyond should be examined for tumourigenicity. The extra population doublings (*e.g.*, 3 to 10) ensure that the results of the tumourigenicity test can be used in the assessment of overall safety of the product even under the assumption of a "worst case" situation and therefore provides a safety buffer.

B.8.4 *Use of control cells*

The tumourigenicity test should include a comparison between the CCL and a positive control reference preparation such as HeLa cells from a reliable source. This source is preferred in order to standardize the test among laboratories, so that the cumulative experience over time can be assessed and made available to NRAs/NCLs and manufacturers to assist them in the interpretation of data. However, other sources for establishing positive-control cells may be acceptable. The purpose of the positive control is to assure that an individual test is valid by demonstrating that the animal model has the capacity to develop tumors from inoculated cells (*i.e.*, a negative result is unlikely to be due to a problem with the *in vivo* model). If the positive-control cells fail to develop tumours at the expected frequency, then this could be indicative of problems with the animals or at the testing facility, such as infections, which can reduce the efficiency of tumour development.

When the cell substrate has been adapted to growth in serum-free medium, which might contain growth factors and other components that could influence growth as well as detection of a tumourigenic phenotype, consideration should be given to processing the positive-control cells in the same medium. Whenever possible, both the test article and the positive-control cells should be resuspended in the same medium, such as phosphate-buffered saline (PBS) for inoculation.

In designing a tumourigenicity protocol, it is important to recognize that tumors arise spontaneously in nude mice and that the incidence of such tumors increases with the age of the mice. Therefore, databases (both published data and the unpublished records/data of the animal production facility that supplied the test animals) of rates of spontaneous neoplastic diseases in nude mice should be taken into account during the assessment of the results of a tumourigenicity test. Generally, negative controls are not recommended because the rates of spontaneous neoplastic disease in nude mice are low, and small numbers of negative control animals are unlikely to provide meaningful data. However, if negative control cells such as WI-38, MRC-5, or FRhl-2 are included, clear justification for including them should be provided. For example, if serum-free medium is used to grow the cell substrate, it is conceivable that growth factors may influence the appearance of spontaneous tumours; therefore, negative-control cells suspended in the same medium may be needed to interpret the test results.

B.8.5 *Number of test animals*

To determine whether the cells being characterized have the capacity to form tumors in animals, the cells being tested, the reference positive-control cells, and if any, the reference negative control cells should be injected into separate groups of 10 animals each. In a valid test, progressively growing tumors should be produced in at least 9 of 10 animals injected with the positive reference cells.

B.8.6 *Number of inoculated cells*

Each animal should be inoculated intramuscularly or subcutaneously [74] with a minimum of 10^7 viable cells. If there is no evidence of a progressively growing nodule at the end of the observation period, the cell line may be considered to be non-tumourigenic. If the cell line is found to be tumourigenic, the NRA/NCL might request additional studies to be done to determine the level of tumourigenicity. This can be done with dose-response studies, where doses of 10^7, 10^5, 10^3 and 10^1 are inoculated, and the data can be expressed as tumor-producing dose at the 50% endpoint (TPD_{50} value) [75].

B.8.7 *Observation period*

Animals are examined weekly by observation and palpation for evidence of nodule formation at the site of injection. The minimum observation period depends on the test system selected. In the case of the nude mouse, a minimum of 4 months is recommended. A shorter period is recommended for the ATS-treated newborn rat, because the immunosuppressive effect of the ATS declines after the final injection at two weeks.

B.8.8 *Assessment of the inoculation site over time (progressive or regressive growth)*

If nodules appear, they are measured in two perpendicular dimensions, the measurements being recorded weekly to determine whether the nodule grows progressively, remains stable, or decreases in size over time. Animals bearing nodules that are progressing should be sacrificed before the end of the study if the tumor reaches the limit set by the relevant authorities for the humane treatment of animals. Animals bearing nodules that appear to be regressing should not be sacrificed until the end of the observation period. Cell lines that produce nodules that fail to grow progressively are not considered to be tumourigenic. If a nodule persists during the observation period and it retains the histopathological characteristics of a tumor, this should be investigated further and discussed with the NRA/NCL.

B.8.9 *Final Assessment of the inoculation site*

At the end of the observation period, all animals, including the reference group(s), are euthanized and examined for gross and microscopic evidence of the growth of inoculated cells at the site of injection and in other sites.

B.8.10 *Evaluation of animals for metastases*

Animals are examined for microscopic evidence of metastatic lesions in sites such as the liver, heart, lungs, spleen, and regional lymph nodes.

B.8.11 *Assessment of metastases (if any)*

Any metastatic lesions are examined further to establish their relationship to the primary tumor at the injection site. If what appears to be a metastatic tumor differs histopathologically from the primary tumor, it is necessary to consider the possibility that this tumor either developed spontaneously or that it was induced by one or more of the components of the cell substrate, such as an oncogenic virus. This may require further testing of the tumor itself or the tumourigenicity assay may need to be repeated. In such cases, appropriate follow-up studies should be discussed and agreed with the NRA/NCL (also see B.9 Oncogenicity).

B.8.12 *Interpretation of results*

A CCL is considered to be tumourigenic if at least 2 of 10 animals develop tumors at the site of inoculation within the observation period. However, the reported rate of spontaneous neoplastic diseases in the test animals should be taken into account during the assessment of the results. In addition, the histopathology of the tumors must be consistent with the inoculated cells, and a genotypic marker should show that the tumor is not of nude mouse origin.

If only one of 10 animals develops a tumor, further investigation is appropriate to determine, for example, if the tumor originated from the cell-substrate inoculum or the host animal and whether there are any viral or inoculated cell DNA sequences present. The NRA/NCL should be consulted in this regard.

The dose-response of the CCL may be studied in a titration of the inoculum as part of the characterization of the CCL. The need for such data will depend upon many factors specific to a given CCL and the product being developed. The NRA/NCL should be consulted in this regard.

B.8.13 *Applicability*

Cell Banks: Representative EOPC or ECB from the MCB or first WBC

Cell Types: DCL, SCL, CCL

B.9 Oncogenicity

B.9.1 *Tests for oncogenicity*

While tumourigenicity is the property of cells to form tumors when inoculated into susceptible animals, oncogenicity is the property of an acellular agent to induce cells of an animal to become tumor cells. As such, tumors that arise in a tumourigenicity assay contain cells derived from the inoculated cells, while tumors that arise in an oncogenicity assay are derived from the host. Oncogenic activity from cell substrates could be due to either the cell substrate DNA (and perhaps other cellular components) or an oncogenic viral agent present in the cells. Although there might be a perception that the cellular DNA from highly tumourigenic cells would have more oncogenic activity than the DNA of weakly or non-tumourigenic cells, at this time, it is not known if there is a relationship between the tumourigenicity of a cell and the oncogenicity of its DNA. Nevertheless, the NRA/NCL might require oncogenicity testing of the DNA and cell lysate from a new cell line (*i.e.*, other than those such as CHO, NS0, Sp2/0, and low passage Vero, for which there is considerable experience) that is tumourigenic in animal model systems (see below) because of the perception that a vaccine manufactured in such a cell line poses a neoplastic risk to vaccine recipients.

The major complication in assaying cellular DNA in animals arises from the size of the mammalian genome. Because the mammalian haploid genome is approximately 3 x 10^9 base pairs (bp) whereas the size of a typical oncogene could be 3-30 x 10^3 bp, the concentration of an oncogene in cellular DNA expression systems would be about 10^5 to 10^6 fold less concentrated than a plasmid containing the same oncogene. As a consequence, if 1 µg of an oncogene expression plasmid induces a tumor in an experimental animal model, the amount of cellular DNA that would contain a similar amount of the same oncogene is 10^5 µg to 10^6 µg (*i.e.*, 100 mg to 1 g). To date, three studies have indicated that between 1 and10 µg of expression plasmids for cellular oncogenes can be oncogenic in mice (33, 42, 43). Therefore, more sensitive *in vivo* assays need to be developed before the testing of the oncogenic activity of cellular DNA becomes practicable. Recent results suggest that the sensitivity of the assay can be increased by several orders of magnitude with the use of certain immune-compromised strains of mice prone to develop tumours after inoculation with oncogenes. Thus, it may be possible to assess the oncogenic activity of cellular DNA in the future. However, at present there is no standardized *in vivo* oncogenicity test for cellular DNA. An example protocol is nonetheless provided in Appendix 3.

Several *in vitro* systems, such as scoring the neoplastic transformation of NIH 3T3 cells in a focus-forming assay following transfection of oncogenic DNA [75,77,78] have been used to assess oncogenicity, but it is not clear how these assays reflect the oncogenic activity of DNA *in vivo*, since they predominantly detect the oncogenic activity of activated *ras*-family members, and thus it is unclear how these assays can assist in estimating risk associated with the DNA or cell lysate from a cell substrate.

Based on experience with DCLs WI-38, MRC-5, and FRhL-2, testing of new MCBs of these cell lines for oncogenicity is not recommended. Other DCLs for which there is substantial experience also may not need to be tested. The NRA/NCL should be consulted in this regard. As stated in Section B.8.1, a new CCL should be presumed to be tumourigenic unless data demonstrate that it is not. If a manufacturer demonstrates that a new CCL is non-tumourigenic, oncogenicity testing on cell DNA and cell lysates might not be required by the NRA/NCL.

When appropriate, and particularly for vaccines, cell DNA and cell lysates should be examined for oncogenicity in a test approved by the NRA/NCL. An oncogenicity testing protocol is provided as Appendix 3.

B.9.2 ***Applicability***
Cell Banks: MCB or first WCB taken to representative EOPC or ECB

Cell Types: CCL, SCL (recommended when tumourigenic cells are used in vaccine production)

B.10 Cytogenetics

B.10.1 *Characterization*

Chromosomal characterization and monitoring were introduced in the 1960s to support the safety and acceptability of human DCLs as substrates for vaccine production. Human DCLs differ from CCLs by retaining the characteristics of normal cells, including the normal human diploid karyotype. A significant quantity of data have been accumulated since then, and this has led to the conclusion that less extensive cytogenetic characterization is appropriate because of the demonstrated karyotypic stability of human DCLs used in vaccine production [79]. Thus, the use of karyology as a lot-by-lot quality-control test is unnecessary for well-characterized and unmodified human DCLs (*e.g.*, WI-38, MRC-5) and for FRhL-2.

Cytogenetic data may be useful for the characterization of CCLs, especially when marker chromosome(s) are identified. Such data might be helpful in assessing the genetic stability of the cell line as it is expanded from the MCB to the WCB and finally to production cultures (see B.3). The following recommendations are appropriate for the characterization of DCL and CCL cell banks.

Cytogenetic recharacterization of DCLs (*e.g.*, WI-38, MRC-5, and FRhL-2) should not be required, unless the cells have been genetically modified or the culture conditions have been changed significantly, since such data are already available (18,19,20). However, for each WCB generated, manufacturers should confirm once that the cells grown in the manner to be used in production are diploid and have the expected lifespan.

For the determination of the general character of a new or previously uncharacterized DCL, samples from the MCB should be examined at approximately four equally spaced intervals during serial cultivation from the MCB through to the proposed *in vitro* cell age used for production or beyond. The testing intervals should be agreed upon with the NRA. Each sample should consist of a minimum of 100 cells in metaphase and should be examined for exact counts of chromosomes, as well as for breaks and other structural abnormalities.

Giemsa-banded karyotypes of an additional five metaphase cells in each of the four samples may provide additional useful information. ISCN [80] 400 band is the minimum acceptable level of Giemsa-banding analysis for human cells.

Stained slide preparations of the chromosomal characterization of the cells (*i.e.*, DCL, CCL), or photographs of these, should be maintained permanently as part of the cell-line record. Further recommendations have been proposed for SCLs [53].

B.10.2 *Applicability*

Cell Banks: MCB, ECB or representative EOPC

Cell Types: DCL, SCL, CCL (as a test for genetic stability, when appropriate),

B.11 Microbial agents

B.11.1 *General considerations*

While many biological production systems require human or animal-cell substrates, such cells are subject to contamination with and have the capacity to propagate extraneous, inadvertent, or so-called adventitious organisms, such as mycoplasma and viral agents. In addition, animal cells contain endogenous agents such as retroviruses that also may be of concern. Testing for both endogenous (*e.g.*, retroviruses) and adventitious agents (*e.g.*, mycoplasmas) is described in the succeeding sections. In general, cell substrates contaminated with microbial agents are not suitable for the production of biological products. However, there are exceptions to this general rule. For example, the CHO and other rodent cell lines that are used for the production of highly purified recombinant proteins express endogenous retroviral particles. Risk *versus* benefit must be considered when determining the suitability of a cell substrate for the production of a specific product. Further, risk-mitigation strategies during production, including purification (removal) and inactivation by physical, enzymatic, and/or chemical means, should be implemented whenever appropriate and feasible. Even though a cell substrate might be unacceptable for some products, such as for a live viral vaccine subjected to neither significant purification nor inactivation, that same cell substrate might be an acceptable choice for a different type of product, such as a highly purified recombinant protein or monoclonal antibody for which risk mitigation has been achieved by significant and validated viral clearance in the production process.

A strategy for testing cell banks for microbial agents should be developed. One strategy is to perform exhaustive testing at the MCB level and to perform more limited testing on the WCB derived from the MCB. This more limited testing would be selected on the basis of those agents that could potentially be introduced during the production of the WCB from the MCB. Testing would not need to be replicated for agents that could only have been present prior to the production of the MCB (*e.g.*, an endogenous retrovirus, or BVDV from serum used for developing the cell seed or in the legacy of establishing the cell line).

However, if the number of vials of a MCB is limited, an alternative strategy would be to conduct the more exhaustive testing on the first WCB made from that MCB, and to limit testing on the MCB itself. An advantage to the strategy of performing more exhaustive testing on the first WCB is that it provides a greater opportunity for amplification of any agents that may have been introduced earlier and through to production of the WCBs. There are advantages and disadvantages to more extensive testing of the MCB or the WCB, and consideration should be given to what is more appropriate for the particular product(s) to be manufactured using a given cell bank. Consultation with the NRA/NCL should be considered prior to implementation to determine whether a proposed testing strategy is acceptable.

EOPC/ECB should be characterized once for each commercial production process. Testing of the ECB serves as further characterization of the MCB or WCB that was exhaustively tested and permits additional time/passages for amplification of low-level contaminants or reactivation of viral contaminants that may have been missed in the testing of the upstream bank.

B.11.2 *Viruses*

Manifestations of viral infections in cell cultures vary widely among the broad array of virus families that are potential contaminants; thus, the methods used to detect them vary. Lytic infections frequently are detected by the CPE they cause. However, in some cases such as non-cytopathic BVDV, no CPE is observed. Viruses also may be present latently (*e.g.*, herpesviruses) or endogenously (in the germline, *e.g.*, retroviral proviruses). Such inapparent infections might require specific techniques designed to reveal their presence, such as molecular and immunological methods, and electron microscopy. For new cell substrates, induction of a detectable infection by exposing the cells to special conditions (*e.g.*, chemical induction; heat shock) may be required, and special detection techniques like transcriptome sequencing or degenerate primer PCR may have utility.

The strategy developed to test cell substrates for viruses should take into consideration the families of viruses and specific viruses that may be present in the cell substrate. Consideration should be given to the species and tissue source from which the cell substrate originated, and to the original donor's medical history in the case of human-derived cell substrates or to the pathogen status of donor animals in the case of animal-derived cell substrates. In addition, consideration should be given to viruses that could contaminate the cell substrate from the donors or from animal- or human-derived raw materials used in the establishment and passage history (legacy) of the cell substrate prior to and during cell banking or production (*e.g.*, serum, trypsin, animal- or human-derived medium components, antibodies used for selection, or animal species through which the cell substrate may have been propagated), as well as laboratory contamination from operators or other cell cultures.

Tests should be undertaken to detect, and where possible identify, any endogenous or exogenous agents that may be present in the cells. Attention should be given to tests for agents known to cause an inapparent infection in the species from which the cells were derived, making it more difficult to detect (*e.g.*, simian virus 40 in rhesus monkeys).

Primary cells are obtained directly from the tissues of healthy animals and are more likely to contain adventitious agents than banked, well-characterized cells. In addition, recent vaccination of source animals should be considered, as they may be exposed to live vaccines. This risk with primary cells can be mitigated by rigorous qualification of source animals and the primary cells themselves. When feasible, animals from which primary cultures are established should be from genetically closed flocks, herds, or colonies monitored for freedom from pathogens of specific concern. Such animals are known as specific-pathogen-free (SPF). The term "closed" refers to the maintenance of a group (flock, herd, or colony) free from introduction of new animals (new genetic material that could introduce, *e.g.*, new retroviral proviruses). Many live viral vaccines are commonly produced in primary cells and undergo little purification during production. In such cases, and when feasible, the use of SPF animals is highly recommended. Documentation of the status of the source animals should be provided to the NRA/NCL. Animals that are not from closed flocks, herds, or colonies should be quarantined and thoroughly evaluated for a period sufficient to detect signs of disease or infection (*e.g.*, monkeys are generally quarantined for 6 or more weeks [81]). Such animals also should be screened serologically for appropriate adventitious agents to determine their suitability as a source for the primary-cell substrate. Animal-husbandry practices should be documented. Even so, viral contamination of the cells may not be excluded from all cultures. For example, contamination of primary monkey kidney cells with foamy virus or simian cytomegalovirus is common in the absence of specific concerted efforts to prevent these contaminations.

For primary-cell cultures, the principles and procedures outlined in Part C, Requirements for Poliomyelitis Vaccine (Oral) [81], together with those in section A.4 of Requirements for Measles, Mumps and Rubella Vaccines and Combined Vaccine (Live) [82] may be followed.

The production of viral vaccines, such as those against smallpox or rabies, originally required the use of living animals, and the great range of possible viral contaminants only became apparent as cell-culture methods were developed. For example, human enteroviruses were not recognized until the development of monkey kidney cell cultures, in which they could produce cytopathic effects, because the disease produced in humans is either relatively mild or in some cases non-existent. It was also clear that viruses could be detected in some systems but not others; for instance, the polyomavirus SV40 does not produce a cytopathic effect in cultures from rhesus monkey kidney cells (in which much of the early polio vaccines were produced) derived from SV40-infected monkeys but will do so in cultures from cynomolgous or African green monkey kidney cells. The suspicion was therefore that there were many viruses in the culture systems of the time and that they were detected only if the assays were appropriate. This remains an accurate view and has lead to a range of different approaches to try and detect all contaminants.

Coxsackie viruses are named after the town in New York where they were first identified and were historically detected by their effects in mice. Coxsackie B viruses produce clinical signs and death in adult mice, while Coxsackie A viruses will affect only suckling mice. For many years, tissue-culture methods were a less reliable method of detection of Coxsackie A viruses than suckling mice, and the continued use of these animals in cell bank characterization reflects this.

In the 1940s, embryonated chicken eggs were a popular substrate for the growth and assay of viruses such as influenza, measles, mumps, yellow fever and vaccinia. They therefore appear to have a wide range of susceptibility. Simian viruses such as SV5 or viruses such as Sendai virus also grow well in them. As many are paramyxoviruses with haemagglutinating (HA) activity, the egg-based assays include tests for HA activity as well as death of the embryos.

A range of tissue-culture cells is also used, typically including one human, one of the same species as the production cell, and one other (often, monkey origin). The hope is that the range will catch viruses not detected by other means, although in practice it is wise to assume that there is no such thing as a generic detection method; for example, cells from an inappropriate monkey species would not necessarily detect SV40 whereas cells from other species will (*e.g.*, Vero cells). In certain circumstances, where a virus is of particular concern, specific tests have been applied. For example, herpes B virus is a common infection of monkeys in the absence of precautions such as quarantine and clinical evaluation of the donor animals, and has very serious effects on infected humans. While herpes B virus was routinely detected by the use of primary rabbit kidney cell cultures, established rabbit cell lines are now acceptable for this purpose. Another example is Marburg virus, which in the 1970s caused a number of deaths in workers who handled monkeys that were to be used in a vaccine-production facility. The incident might have been avoided had the animals been adequately quarantined. A specific test in guinea pigs was introduced and maintained for a number of years to ensure the absence of the agent.

There is a disparate range of tests that have been or are still used with the aim of detecting any significant contaminant that may be present in cell cultures. Some, such as the rabbit kidney cell test, are very specific in intent while others may be expected to be more generic. In general, however, it is wise to assume that an assay will never be all encompassing whether based on historical virological approaches or more current methodologies. Thus, the consequences of deleting tests on the grounds of redundancy must be very carefully evaluated before any action is taken. On the other hand, it is difficult to justify maintenance of a test if it only detects viruses also detected by other methods of equivalent sensitivity, comparable ease of use, and cost. Each of these considerations should be given when developing an appropriate testing strategy for the given cell bank. Policies to minimize the use of animals in safety testing should also be considered, but must be considered in balance with the utility and necessity (sensitivity and ability to detect particular adventitious agents not readily detected by other means) of the test in which they are used.

B.11.2.1 *Tests in animals and eggs*

The cells of the MCB and WCB are unsuitable for production if any of the animal or egg tests shows evidence of the presence of any viral agent attributable to the cell banks.

Generally, MCBs are thoroughly characterized by the methods listed below. WCBs may be characterized by an abbreviated strategy, when appropriate. However, an alternative strategy to this general rule may be used, as discussed in section B.11.2 above. These tests may be performed directly on cells or supernatant fluids or cell lysates from the bank itself or on cells or supernatant fluids or cell lysates from passaged cells from the bank that have been passaged to the proposed *in vitro* cell age for production or beyond.

In some countries, policies exist to minimize the use of animals in safety testing.

B.11.2.1.1 Adult mice

The original purpose of this test was for the detection of lymphocytic choriomeningitis virus (LCMV). The test in adult mice for pathogenic viruses includes inoculation by the intraperitoneal route (0.5 mL) with cells and culture fluids from the MCB or WCB, where at least 10^7 viable cells or the equivalent cell lysate are divided equally among at least ten adult mice weighing 15-20 g.

In some countries, the adult mice are also inoculated by the intracerebral route (0.03 mL).

In some countries at least 20 mice are required for each test.

The animals are observed for at least 4 weeks. Any animals that are sick or show any abnormality are investigated to establish the cause. Animals that do not survive the observation period should be examined for gross pathology and histopathology in order to determine a cause of death, and if a viral infection is indicated, efforts should be undertaken to identify the virus. Viral identification may involve culture and/or molecular methods. Further, each mouse that dies after the first 24 hours of the test, or is sacrificed because of illness, should be necropsied and examined for evidence of viral infection by subinoculation of appropriate tissue into at least five additional mice which should be observed for 21 days. The test is not valid if more than 20% of the animals in either or both of the test and negative control groups (if used) become sick for non-specific reasons and do not survive the observation period.

In some countries, the adult mice are observed for 21 days.

If the cell substrate is of rodent origin, at least 10^6 viable cells or the equivalent cell lysate are injected intracerebrally into each of ten susceptible adult mice to test for the presence of LCMV.

In some countries, after the observation period, the animals are challenged with live LCMV to reveal the development of immunity against non-pathogenic LCMV contaminants resulting in otherwise unapparent infection.

B.11.2.1.1.1 Applicability

Cell Banks: MCB, WCB, or ECB or representative EOPC

Cell Types: PCC, DCL, SCL, CCL

B.11.2.1.2 Suckling mice

The original purpose of this test was for the detection of Coxsackie viruses. The test in suckling mice for pathogenic viruses includes inoculation by the intraperitoneal route (0.1 mL) with cells and culture fluids from the MCB or WCB; at least 10^7 viable cells or the equivalent cell lysate are divided equally between two litters of suckling mice, comprising a total of at least ten animals less than 24-hours old.

In some countries, the suckling mice are also inoculated by the intracerebral route (0.01 mL).

In some countries, 20 suckling mice are inoculated.

The animals are observed for at least 4 weeks. Any animals that are sick or show any abnormality are investigated to establish the cause. Animals that do not survive the observation period should be examined for gross pathology and histopathology in order to determine a cause of death, and if a viral infection is indicated, efforts should be undertaken to identify the virus, where this is practicable. Viral identification may involve culture and/or molecular methods. Further, such examination of viral infection should include subinoculation of appropriate tissue suspensions into an additional group of at least five suckling mice by intracerebral and intraperitoneal routes and observation daily for 14 days. In the case of suckling mice, it is often observed that those that perish are cannibalized by their mother and this renders determining a cause of death impossible (when they are fully cannibalized and no remains can be recovered). The test is not valid if more than 20% of the animals in either or both of the test group and negative control group (if used) do not survive the observation period.

In some countries, the suckling mice may be observed for a period of 14 days followed by a sub-passage involving a blind passage (*via* intraperitoneal and intracerebral inoculation into at least 5 additional mice) of a single pool of the emulsified tissue (minus skin and viscera) of all mice surviving the original 14-day test.

B.11.2.1.2.1 Applicability

Cell Banks: MCB, WCB or ECB or representative EOPC

Cell Types: PCC, DCL, SCL, CCL

B.11.2.1.3 Guinea pigs

The original purpose of this test was for the detection of LCMV and Mycobacterium tuberculosis. When it is necessary to detect Mycobacterium species, a test in guinea pigs is performed and includes inoculation by the intraperitoneal route (5 mL) with cells and culture fluids from the MCB or WCB, where at least 10^7 viable cells or the equivalent cell lysate are divided equally among the animals.

In some countries, tests in five guinea-pigs weighing 350-450 g are also inoculated by the intracerebral route (0.1 mL) and observed for 42 days to reveal *Mycobacterium tuberculosis* and other species.

The animals are observed for at least 6 weeks. Any animals that are sick or show any abnormality are investigated to establish the cause. Animals that do not survive the observation period should be examined for gross pathology and histopathology in order to determine a cause of death, and if a viral infection is indicated, efforts should be undertaken to identify the virus. Viral identification may involve culture and/or molecular methods. The test is not valid if more than 20% of the animals in either or both of the test and negative control (if used) groups do not survive the observation period.

The test in guinea-pigs for the presence of Mycobacterium may be replaced by an alternative *in vitro* method such as culture, or shortened culture with a PCR endpoint (also see B.11.3).

B.11.2.1.3.1 Applicability

Cell Banks: MCB, WCB or ECB or representative EOPC

Cell Types: PCC, DCL, SCL, CCL (the latter three are dependent on legacy and current use of media components of animal origin that could result in contamination with Mycobacterial species)

B.11.2.1.4 Rabbits

The original purpose of this test was for the detection of Herpes B virus. When it is necessary to detect simian herpes B virus, the test in rabbits for pathogenic viruses is performed and includes the inoculation by the intradermal (1 mL) and subcutaneous (>2 mL) routes with cells and culture fluids from the MCB or WCB, where at least 10^7 viable cells or the equivalent cell lysate are divided equally among the animals.

In some countries, tests in five rabbits weighing 1.5-2.5 kg are inoculated by the subcutaneous route with either 2 mL or between 9 and 19 mL. Consultation with the NRA/NCL regarding acceptable methods should be considered.

The animals are observed for at least 4 weeks. Any animals that are sick or show any abnormality are investigated to establish the cause. Animals that do not survive the observation period should be examined for gross pathology and histopathology in order to determine a cause of death, and if a viral infection is indicated, efforts should be undertaken to identify the virus. Viral identification may involve culture and/or molecular methods. The test is not valid if more than 20% of the animals in either the test or the negative control (if used) groups do not survive the observation period.

The test in rabbits for the presence of herpes B virus is intended for primary simian cultures, and may be replaced by a test in rabbit kidney cell cultures.

B.11.2.1.4.1 Applicability

Cell Banks: MCB, WCB or ECB or representative EOPC

Cell Types: PCC, DCL, SCL, CCL

B.11.2.1.5 Embryonated chicken eggs

At least 10^6 viable cells or the equivalent cell lysate, along with culture fluids, from the MCB or WCB of avian origin, propagated to the proposed *in vitro* cell age for production or beyond are injected into the allantoic cavity of each of at least ten embryonated hens' eggs, and into the yolk sac of each of at least another ten embryonated hens' eggs. The eggs are examined after not fewer than 5 days of incubation. The allantoic fluids of the eggs are tested with red cells from guinea pig and chickens (or other avian species) for the presence of haemagglutinins. The test is not valid if more than 20% of the embryonated hens' eggs in either or both of the test group and negative control group (if used) are discarded for non-specific reasons.

In some countries, the NRA/NCL also requires that other types of red cells, including cells from humans (blood group IV O) or monkeys should be used in addition to guinea-pig and chicken (or other avian species) cells. In all tests, readings should be taken after incubation for 30 minutes at 0-4°C, and again after a further incubation for 30 minutes at 20-25°C. For the test with monkey red cells, readings also should be taken after a final incubation for 30 minutes at 34-37°C.

In some countries, inoculation by the amniotic route is used.

In some countries, following incubation, allantoic fluids or a 10% suspension of yolk sacs, as appropriate, should be harvested, pooled, and blind passaged into an additional group of eggs.

Usually, the eggs used for the yolk sac test should be 5-7 days old. The eggs used for the allantoic cavity test should be 9-11 days old.

Alternative ages for the embryonated chicken eggs and alternative incubation periods are acceptable if they have been determined to be equivalent or better at detecting the presence in the test samples of the adventitious agents this test is capable of detecting when performed as above.

Embryos that do not survive the observation period should be examined for gross pathology in order to determine a cause of death, and if a viral infection is indicated, efforts should be undertaken to identify the virus. Viral identification may involve culture and/or molecular methods.

B.11.2.1.5.1 Applicability

Cell Banks: MCB of avian origin, WCB of avian origin or ECB or representative EOPC

Cell Types: avian PCC, DCL, SCL, CCL (also recommended for novel cell substrates)

B.11.2.1.6 Antibody-production tests

Rodent cell lines are tested for species-specific viruses using mouse, rat and hamster antibody production tests, as appropriate. *In vivo* testing for lymphocytic choriomeningitis virus, including a challenge for non-lethal strains, is performed for such cell lines, as described in Section B.11.2.1.1. Avian cell lines may also be tested using a chick antibody-production test, *e.g.*, to detect chicken anemia virus. Further, if the cell substrate (even if not of rodent origin) has been exposed to rodent-origin materials, *e.g.*, selection using a monoclonal antibody, testing should be considered for the relevant-species viruses using an antibody-production test [83, 84].

In some countries, consideration is being given to use of nucleic acid testing (NAT) in place of the *in vivo* antibody-production testing. In these cases, data should be provided to the NRA/NCL that justifies this practice.

B.11.2.1.6.1 Applicability

Cell Banks: MCB, WCB, or ECB or representative EOPC

Cell Types: DCL, SCL, CCL (recommended primarily for cells of rodent origin)

B.11.2.2 *Tests in cell culture*

Tests in cell culture are capable of detecting a broad array of viral families. Readouts include monitoring the cultures periodically for CPE and tests for haemadsorbing and haemagglutinating viruses, which are conducted at the end of the culture period. In addition to the indicator cells described below, it may be appropriate to expand the different types of indicator cells used (beyond 2 or 3) to enable the detection of viruses with differing host requirements. Decisions about which cell lines to use as indicator cells should be guided by the species and legacy of the production cell substrate taking into the consideration the types of viruses to which the cell substrate could potentially have been exposed and thus the viruses one would like to detect by this assay method.

The cell substrate is unsuitable for production if any of the indicator cell cultures shows evidence of the presence of any viral agent attributable to the tested cell substrate.

B.11.2.2.1 Applicability

Cell Banks: MCB, WCB or ECB or representative EOPC

Cell Types: PCC, DCL, SCL, CCL

B.11.2.2.2 Indicator cells

Live cells or cell lysate, each with spent culture fluids of the MCB or WCB are inoculated onto monolayer cultures or cultivated with monolayer cultures of the cell types listed below, as appropriate.

A lysate of the cells may be prepared by a method that avoids virus disruption while allowing maximal virus release (*e.g.*, typically three freeze/thaw cycles followed by low-speed centrifugation). If cells, lysate, or spent culture fluids are to be stored prior to testing, then they should be stored at $\leq$-70°C.

- Cultures (primary cells or CCL) of the same species and tissue type as that used for production.
- Cultures of a human DCL. The original purpose of this test, utilizing primary human cells, was the detection of measles virus. But, where the cell substrate is of human origin, a simian kidney cell line should be used as the second indicator cell line. The original purpose of the use of this cell type was the detection of simian viruses.

In some countries, cultures of another (third) cell line from a different species are required.

In many circumstances, more than two cell lines may be necessary to cover the range of potential viral contaminants and typically, a third cell line would be used that is of simian origin, if the cell substrate is not of simian origin.

For new cell substrates, additional cell lines to detect viruses known to be potentially harmful to humans could be considered (*e.g.*, for insect cell lines, if the cells selected for the above mentioned tests are not known to be permissive to insect viruses, an additional detector cell line should be included in the testing).

The cell bank sample to be tested is diluted as little as possible. At least 10^7 cells, or equivalent cell lysate, and spent culture fluids are inoculated onto each of the indicator cell types. The resulting co-cultivated or inoculated cell cultures are observed for evidence of viruses by cytopathic effect for at least 2 weeks. If the cell line is known to be capable of supporting the growth of human or simian cytomegalovirus, HDC cultures are observed for at least 4 weeks. Extended (4-weeks) cell culture for the purposes of detecting human or simian cytomegalovirus can be replaced by the use of NAAT to detect cytomegalovirus nucleic acid.

In some countries, a passage onto fresh cultures for an additional 2 weeks is recommended for all indicator cultures. In some cases, it may be difficult to keep the cell cultures healthy for 2 weeks without subculturing. In those cases, it may be necessary to feed the cultures with fresh medium or to subculture after two weeks onto fresh cultures in order to be able to detect viral agents.

At the end of the observation period, samples of each of the co-cultivated or inoculated cell culture systems are tested for haemadsorbing and/or haemagglutinating viruses, as described in section B.11.2.1. 5.

B.11.2.2.3 Additional considerations to the tests in cell culture for insect viruses

Many insect cell lines carry persistent viral infections that do not routinely produce a noticeable CPE (*e.g.*, some clones of the Hi-5 cell line are persistently infected with an insect nodavirus). However, the viruses may be induced to replicate by stressing the cells using a variety of techniques such as increased/reduced culture temperature (above or below that routinely used for production), heat-shock for a short period, super-infection with other insect viruses, or chemical inducers. Therefore, the probability of detecting such low-level persistent infections may be increased by stressing the cells prior to analysis.

Intact cells and cell lysates from a passage level at or beyond that equivalent to the EOPC is co-cultivated with indicator cells from at least three different species of insect in addition to the indicator cells as noted in section B.11.2.2.2. Cell lines should be selected on the following basis: one of the lines has been demonstrated to be permissive for the growth of human arboviruses, one has been shown to be permissive for the growth of a range of insect viruses, and the third has been derived from a species that is closely related to the host from which the MCB is derived (or another line from the same species). Duplicate cultures of indicator cells are typically incubated at two temperatures, such as 37 ± 1°C and a lower temperature, such as 28 ± 1°C, observed for a period of 14 days, and examined for possible morphological changes. The cell-culture fluids from the end of the test period are tested for haemagglutinating viruses, or the intact cells from the end of the test period are tested for haemadsorbing viruses. The cells comply with the test if no evidence of any viral agent is found.

Several mosquito cell lines are available that are permissive for the growth of some human arboviruses and could be considered for these tests. Alternatively, BHK-21 cells could be considered for this purpose. The most permissive insect cell lines characterized to date have been derived from embryonic Drosophila tissues. While the mosquito and Drosophila cell lines may be suitable for some aspects of the testing, it should be remembered that many insect cell lines are persistently infected with insect viruses that usually produce no obvious CPE. In addition, many insect cells may be infected with mammalian viruses, such as BVDV, that are known to replicate in insect cells. Demonstrating that the indicator cell lines are themselves free from adventitious agents is an important pre-requisite to their use in the testing outlined above. Consideration should also be given to risk-mitigation strategies as discussed above for highly purified products for which viral clearance can be achieved and validated.

B.11.2.3 *Transmission electron microscopy*

At least 200 cells from the MCB or WCB, and ECB are examined by transmission electron microscopy (TEM) for evidence of contamination with microbial agents. Methods include negative staining and thin section. A discussion of these methods is provided by Bierley *et al.* [85]. It may be appropriate to examine more cells in some cases, such as discussed below for insect cell lines, and the NRA/NCL should be consulted in this regard. Any unusual or equivocal observations that might be of microbiological significance should be noted and discussed with the NRA/NCL.

TEM can detect viral particles in a cell substrate, including certain endogenous retroviruses. While TEM is fairly insensitive (generally detecting gross contamination, but not necessarily low-level contamination), it is a generic assay that can detect microbial agents of many types. –

B.11.2.3.1 Applicability

Cell Banks: MCB, WCB, or ECB or representative EOPC

Cell Types: DCL, SCL, CCL

B.11.2.3.2 Additional considerations to TEM for insect cells

For MCBs and WCBs derived from insect cells, the general screening test outlined above applies. In addition, cell lines should be subjected to stress conditions, such as described in section B.11.2.2.3, prior to examination by transmission electron microscopy. Further, increasing the number of cells examined may also improve the probability of detecting an agent (*e.g.*, errantiviruses and hemiviruses). The maintenance temperatures and treatments used should be agreed with the NRA/NCL along with the number of sectioned cells to be examined.

B.11.2.4 *Tests for retroviruses*

All vertebrate and insect cells that have been analyzed possess endogenous, genetically acquired retroviral sequences integrated into chromosomal DNA in the form of proviruses. These sequences may be expressed, or be induced, as mRNA. In some cases, the mRNA is translated into viral protein and virus particles (virions) are produced. In many cases, these virions are defective for replication (*e.g.*, avian endogenous retrovirus EAV, Chinese hamster ovary cell line gamma-retrovirus [86]) whereas in others (*e.g.*, X-MuLV) the retroviruses may be capable of infecting cells of other species including human cells.

Consideration should also be given to the possibility that cell banks may be infected with non-genetically acquired retroviruses (exogenous retroviruses), either because the donor animal was infected or through laboratory contamination.

It should be noted that infection by retroviruses is not necessarily associated with any cytopathic effect on the cells and therefore it may require screening assays, like the Product-Enhanced RT (PERT) assay for reverse transcriptase or TEM to reveal their presence.

The cells of the MCB or WCB are unsuitable for production if the tests for infectious retroviruses, if required, show evidence of the presence of any viral agent attributable to the substrate that cannot be demonstrated to be cleared during processing. Generally, the downstream manufacturing process for products (*e.g.*, monoclonal antibodies) made in cell substrates that produce retroviral particles (*e.g.*, CHO cells) or infectious endogenous retrovirus (*i.e.*, NS0, SP2/0 cells) is validated to provide adequate viral clearance [14]. The margin of viral clearance required should be agreed with the NRA/NCL.

Chick embryo fibroblasts (CEF) contain defective retroviral elements that frequently produce defective particles with reverse transcriptase activity. This has been the subject of many studies and WHO consultations because they are used for live viral vaccine production. If evidence is presented that the donor flock is free of infectious retroviruses and there is no evidence that the cultures are contaminated with infectious retroviruses, then the cultures can be considered acceptable with respect to retrovirus tests.

Rodent cell lines express endogenous retroviruses, and thus infectivity tests should be performed to determine whether these endogenous retroviral particles are infectious.

Cell lines such as CHO, BHK-21, NS0, and Sp2/0 have frequently been used as substrates for drug production with no reported safety problems related to virus contamination of the products and may be classified as "well-characterized" because the endogenous retrovirus particles have been studied extensively. Furthermore, the total number of retrovirus-like particles present in the harvest is evaluated quantitatively (TEM or QPCR) on a representative number of lots and retrovirus clearance is demonstrated with significant safety factors. Thus, in these situations testing for infectious retrovirus may be reduced; *e.g.*, test one lot then discontinue testing, but repeat when there is a significant change in the cell-culture process such as a change in scale. Sponsors are encouraged to consult with the NRA.

B.11.2.4.1 Applicability

MCBs and cells that have been propagated to the proposed *in vitro* cell age for production or beyond. Alternatively, this testing could be performed on WCBs.

Cell Banks: MCB or WCB, and ECB or representative EOPC

Cell Types: DCL, SCL, CCL

B.11.2.4.2 Reverse Transcriptase (RT) Assay

Test samples from the MCBs or WCBs propagated to the proposed *in vitro* cell age for production or beyond are examined for the presence of retroviruses.

Culture supernatants are tested by a highly sensitive, quantitative, polymerase chain reaction (PCR)-based RT assay or PERT assay [87,88,89,90].

RTase activity is not specific to retroviruses and may derive from other sources, such as retrovirus-like elements that do not encode a complete or infectious genome [91,92,93,94,95] or cellular DNA-dependent DNA polymerases [96,97]. Attempts to reduce the PERT activity associated with cellular DNA-dependent DNA polymerases have been reported [97,98,99], although no treatment can eliminate all activity, and thus the results of such highly sensitive assays need to be interpreted with caution. Use of appropriate controls in the assay assist in this regard. Since RT activity can be associated with the presence of defective retrovirus-like particles and since polymerases other than reverse transcriptase can result in apparent RT activity, a positive result in an RT assay is not conclusive evidence of the presence of infective retrovirus. Positive results may require further investigation, *e.g.*, carrying out infectivity assays (see Section B.9.1.4.4). Also, it may be useful to utilize the conventional RT assay in this investigation to determine whether the RT activity is Mg++ or Mn++ dependent. Such testing should be agreed in advance with the NRA/NCL.

CEFs and other cells of avian origin are known to express retroviral elements. With such cells, the appropriateness of this test should be discussed with the NRA/NCL. For example, it may be appropriate to direct testing strategies at the detection of infectious avian retroviruses, such as avian leukosis viruses and reticuloendotheliosis virus, including serological screening of flocks that are the source of the CEFs. Additionally, it is known that insect cells have retroviral elements that are detected by a PERT assay, and so they too can test positive by this assay.

B.11.2.4.3 PCR or other specific *in vitro* tests for retroviruses

When the PERT test gives unclear results, or when it is unavailable, it may be appropriate to screen the cell substrate for species-specific retroviruses by molecular methods, such as the PCR, immunofluorescence, ELISA, or other virus-specific detection methods. Molecular methods, such as PCR, also may be used for quantification of retrovirus like particles in the production harvests provided that the method is validated accordingly. Consultation with the NRA/NCL regarding the acceptability of this approach is recommended.

B.11.2.4.4 Infectivity test for retroviruses

When the test sample is found to have RT activity, it might be necessary to carry out infectivity assays to assess whether the activity is associated with replicating virus.

Because rodent cells generally express endogenous retroviruses, the infectivity and *in vitro* host range of such retroviruses should be assessed. Test samples from the MCB or WCB, propagated to the proposed *in vitro* cell age for production or beyond should be examined for the presence of retroviruses by infectivity assays. Cells to be used for these assays should be able to support the replication of a broad range of viruses; this might require using cells of various species and cell types. The testing strategy should be agreed with the NRA/NCL.

It is often possible to increase the sensitivity of assays by first inoculating the test material onto cell cultures that can support retroviral growth in order to amplify any retrovirus contaminant that may be present at low concentrations. For non-murine retroviruses, test cell lines should be selected for their capacity to support the growth of a broad range of retroviruses, including viruses of human and non-human primate origin [100,101].

For murine retroviruses, it is important to assess whether the cells release infectious retroviruses and, if so, to determine the host range of those viruses. The testing for murine retroviruses can be complex, and the NRA/NCL should be consulted for guidance. Murine and other rodent cell lines (CHO, NS0, SP2/0) or hybrid cell lines containing a rodent component should be assumed to be inherently capable of producing infectious retroviruses or non-infectious retrovirus-like particles. In such cases, the clearance (removal and/or inactivation) of such retroviruses during the manufacturing process should be quantified and provide a level of clearance acceptable to the NRA/NCL.

Any testing proposed by the manufacturer should be agreed with the NRA/NCL.

B.11.2.5 *Tests for particular viruses not readily detected by the tests described in sections B.11.2.1, B.11.2.2, B.11.2.3, and B.11.2.4 and their sub-sections*

Some viruses, such as hepatitis B or C viruses or human papillomaviruses, cannot be detected readily by any of the methods described above because they are not known to grow readily in cell culture or are restricted to human host range. Some animal viruses, *e.g.*, bovine polyomavirus and porcine circoviruses, are not readily detected by the routine tests previously described. In such circumstances, it may be necessary to include specific assays for such viruses. While broad general tests are preferable for detecting unknown contaminants, some selected viruses may be screened by using specific assays, such as molecular techniques (*e.g.*, nucleic acid amplification). Antibody-based techniques might also be employed, such as immunofluorescence assays.

Generally, once the MCB, WCB, or ECB has been demonstrated to be free of selected viruses, it might not be necessary to test the cells at later stages (*e.g.*, at the production level) if such viruses could not be introduced readily during culture.

Human cell lines should be screened using appropriate *in vitro* techniques for specific viruses that are the cause of significant morbidity, for those viruses that might establish latent or persistent infections and for viruses that may be difficult or impossible to detect by the techniques described in sections B.11.2.1, B.11.2.2, B.11.2.3, and B.11.2.4 and their sub-sections. Selection of the viruses to be screened should take into account the tissue source and medical history of the donor, if available, from whom the cell line was derived.

Under circumstances in which the cell origin or medical history of the donor, if available, would suggest their presence, it may be appropriate to perform specific testing for the presence of human herpesviruses, human retroviruses, human papillomaviruses, human hepatitis viruses, human polyomaviruses, or difficult-to-culture types of human adenoviruses.

Consideration should be given to screening insect cell lines for specific viruses that have been reported to contaminate that particular cell line (*e.g.*, nodaviruses) or viruses that may be present persistently in insect cell lines and that are known to be infectious for humans.

B.11.2.5.1 Applicability

The NRA/NCL should be consulted in regard to the specific pathogens or selected viruses that should be included in the testing strategy as these will be directed on a case-by-case basis depending on the species and origin of the cell and the medical history of the donor, if available.

Cell Banks: MCB, WCB, or ECB or representative EOPC

Cell Types: PCC (as needed), DCL, SCL, CCL

B.11.2.5.2 Nucleic acid detection methods

Tests for selected viruses usually are performed using nucleic acid amplification and detection methods. PCR can be performed directly on DNA extracted from the cells or on cell lysates or supernatant fluids by DNA amplification, or on RNA by reverse transcription followed by DNA amplification (RT-PCR). In this manner, both DNA and RNA viruses can be detected, as can the proviral DNA of retroviruses. PCR primers can be directed against variable regions of viral nucleic acids in order to ensure detection of a specific virus or viral strain, or against conserved regions of viral sequences shared amongst strains or within a family in order to increase the opportunity of detecting multiple related viruses. Standard PCR analysis can be coupled with hybridization methods to increase its versatility, sensitivity, and specificity. For example, the use of probes to various regions of the amplicon might be useful in identifying the virus strain or family. However, PCR methods have the limitation that viral genes might not be sufficiently conserved among all members of any particular viral family to be detected even when conserved regions are selected.

New, sensitive, molecular methods, with broad detection capabilities are being developed. These are not yet in routine use, but as they become widely available and validated, they will play an increasing role in the evaluation of cell substrates. The sensitivity of these methods as well as their breadth of detection should be considered when evaluating their applicability. One of the advantages of some of these new methods is that they have the potential to discover new viruses. These new approaches involve either degenerate PCR for whole virus families or random-priming methods, which do not depend on a known sequence. Analysis of the resulting amplicons has employed sequencing, hybridization to oligonucleotide arrays, and mass spectrometry [102, 103, 104]. The new generation of massively parallel (deep) sequencing (MPS) methods may have particular utility. They can be applied to detect virions after nuclease treatment to remove cellular DNA and unencapsidated genomes. Used in this mode, MPS has been used to discover new viruses in serum and other tissues and has revealed the contamination of human vaccines by porcine circovirus [102, 105, 106, 107, 108, 109]. MPS can also be employed to screen cell substrates for both latent and lytic viruses by sequencing the transcriptome. In this mode, enormous quantities of data are generated, and robust bioinformatic methods are required to detect viral sequences by either positive selection against viral databases or negative selection to remove cellular sequences [102, 109, 110]. Care is required to exclude false "hits" to viruses due to recognition of transduced cellular sequences present in some viral genomes or due to viral genes like virokines that have a close homology to cellular genes.[102,104, 110]

It is probable that application of methods of this type will be expected or required by regulatory agencies in future. At present they have not been evaluated for sensitivity and specificity and should be thought of as powerful investigational tools that can reveal issues to be explored by more established methods.

B.11.3 *Bacteria, fungi, mollicutes, and mycobacteria*

The most common contaminants of cell culture are non-viral. These can be introduced easily from the environment, materials, personnel, *etc.* Furthermore, many such organisms multiply rapidly and can be pathogenic for humans. It is also important in risk evaluation for the manufacturer to bear in mind that standard compendial tests for "sterility" are intended to give an indication of the effectiveness of aseptic processing to prevent general bacterial or fungal contamination and are not capable of isolating all potential bacterial and fungal contaminants. The manufacturer should consult with the NRA/NCL regarding any particular materials or environments where there may be an elevated hazard of contamination with particular types of fastidious organisms.

Biological starting materials, like cell substrates, should be characterized to ensure that they are free from adventitious infectious organisms such as bacteria, fungi, cultivable and non-cultivable mycoplasmas, spiroplasma (in the case of insect cells or cells exposed to plant-derived materials), and mycobacteria. For a substance to be considered free of such contaminants, the assays should demonstrate, at a predefined level of sensitivity, that a certain quantity of the substance does not contain detectable levels of that contaminant. Testing should be conducted in an aseptic environment under appropriate clean room conditions to avoid false-positive results. Testing should include a plan to account for the need for repeat testing to deal with potentially false-positive results and a prequalification plan for reagents used in the tests.

Mycobacterial testing might be applied to cell-bank characterization if the cells are susceptible to infection with *Mycobacterium tuberculosis* or other species. Such testing also should be performed on primary-cell cultures. It may be necessary to lyse the host cells in order to detect Mycobacteria, because some strains may be primarily intracellular.

Detection of mycoplasma/spiroplasma may require different growth conditions from methods used for mammalian cells - although at least one, Spiroplasma, can be cultivated at 30ºC. Positive controls for these tests (particularly for Spiroplasmas) are an issue that needs to be resolved. Spiroplasma have been reported as infectious agents in a number of insect species and insect cell lines, and in addition have been reported to cause pathogenic effects in mammals.

B.11.3.1 *Bacterial and Fungal Sterility*

Tests are performed as specified in Part A, section 5.2 [111] of the Requirements for Biological Substances No. 6 (General Requirements for the Sterility of Biological Substances), by a method approved by the NRA/NCL. Additional information can be found in national pharmacopoeias and ICH documents [10,112,113,114]. For the MCB and WCB, the test is carried out using for each medium 10 mL of supernatant fluid from cell cultures. In addition, the test is carried out on at least 1% of the filled containers (*i.e.*, cyropreservation vials) with a minimum of two containers. For supernatant fluid, the recommended method is the membrane filtration method. For cell bank vial testing, it may be necessary to use the direct inoculation method. Bacteriostasis and fungistasis should be excluded.

B.11.3.1.1 Applicability

Cell Banks: MCB, and each WCB

Cell Types: PCC, DCL, SCL, CCL

B.11.3.2 *Mollicutes*

Mollicutes are distinguished by an absence of a cell wall and includes mycoplasmas, acholeplasmas, spiroplasmas, and others. They are parasites of various animals and plants, living on or in the host's cells. They are also a frequent contaminant of cell cultures. In addition to their potential pathogenicity, mycoplasmas compete for nutrients, induce chromosomal abnormalities, interrupt metabolism and inhibit cell fusion of host cells. M. pneumoniae is pathogenic for humans, although there are no reported cases of human infections with this organism arising from exposure to cell cultures or cell-derived products. In any case, cell banks should be demonstrated to be free of such contamination in order to be suitable for the production of biologicals.

B.11.3.2.1 Mycoplasma and acholesplasma

Tests for mycoplasmas are performed as specified in Part A, sections 5.2 and 5.3 of the Requirements for Biological Substances No. 6 (General Requirements for the Sterility of Biological Substances) [115], or by a method approved by the NRA/NCL. Both the culture method and the indicator cell-culture method should be used. NAAT alone, in combination with cell culture, or with an appropriate detection method, might be used as an alternative to one or both of the other methods after suitable validation and discussion with the NRA/NCL. In this case, a comparability study should be carried out. The comparability study should include a comparison of the respective detection limits of the alternative method and official methods. Specificity (mycoplasma panel detected, potential false-positive results due to cross-reaction to other bacteria) should also be considered. More details are available in the European Pharmacopoeia chapter 2.6.7 [116]. One or more containers of the MCB and each WCB are used for the test.

B.11.3.2.1.1 Applicability

Cell Banks: MCB and each WCB

Cell Types: PCC, DCL, SCL, CCL

B.11.3.2.2 Spiroplasma and others

Other mollicutes such as Spiroplasma may be introduced in cell substrates through contamination of raw materials (Peptons) or due to the nature and permissivity of the cells (*e.g.*, insect cells). According to the cell bank manufacturing process, *e.g.,* if the raw material exposure is at the level of MCB or before, it may be appropriate to test the MCB only. If further exposure is possible, then testing of WCB might be necessary too.

Detection of such mollicutes may require adapted culture conditions (medium and/or temperature) depending on the strain to be detected. To guarantee a broad detection of such mollicutes, it is helpful to use NAAT after suitable validation with an appropriate model (*e.g.*, *Spiroplasma citri* or other strain according to the cell origin).

B.11.3.2.2.1 Applicability

Cell Banks: MCB, WCB (recommended for insect cells)

Cell Types: DCL, SCL, CCL (recommended for insect cell substrates and when raw materials of plant origin are used during the cell bank preparation or production process).

B.11.3.3 Mycobacteria

The test for mycobacteria is performed as described below or by a method approved by the NRA/NCL.

Inoculate 0.2 mL of the sample in triplicate onto each of 2 suitable solid media (such as Löwenstein-Jensen medium and Middlebrook 7H10 medium). Inoculate 0.5 mL in triplicate into a suitable liquid medium at 37°C for 56 days.

In some countries, the incubation period is 42 days.

An appropriate positive control test should be conducted simultaneously with the sample under evaluation, and the test should be shown to be capable of detecting the growth of small amounts of mycobacteria. In addition, the fertility of the medium in the presence of the preparation to be examined should be established by a spiking inoculation of a suitable strain of a Mycobacterium sp., such as BCG. If at the end of the incubation time no growth of mycobacteria occurs in any of the test media and the positive control and spiked control show appropriate growth, the preparation complies with the test.

Nucleic acid amplification techniques might be used as an alternative to this culture method, provided that such an assay is shown to be comparable to the compendial culture method. An appropriate comparability study should be carried out that includes a comparison of the respective detection limits of the alternative method and culture method. Specificity (mycobacteria panel detected, potential false-positive results due to cross-reaction to other bacteria) should also be considered. An *in vivo* method, as described in the test in guinea pigs, also may be used (B.11.2.1.3).

B.11.3.3.1 Applicability

Cell Banks: MCB or WCB

Cell Types: PCC, DCL, SCL, CCL

B.11.4 *Transmissible Spongiform Encephalopathies*

TSEs are a group of slowly developing fatal neurological diseases affecting the brain of animals and humans. The accepted view at present is that they are caused by non-conventional infectious agents known as prions (PrP^{tse}), which are made up of a normal host protein (PrP) in an abnormal conformation. TSEs include BSE of cattle, scrapie of sheep, CJD and its variant form (vCJD), GSS and FFI in humans, CWD in elk and deer, and transmissible mink encephalopathy (TME) [117,118]. Normal PrP (PrPc) protein may be expressed on cell surfaces, but *in vivo* this protein can mis-fold and become the abnormal disease-causing type PrP^{tse}, which is able to catalyze the conversion of PrPc protein into the abnormal conformation. PrP^{tse} is relatively resistant to common proteolytic enzymes, such as proteinase K, compared with PrPc.

Bovine spongiform encephalopathy (BSE) was first described in the United Kingdom in 1984, and the numbers of clinical cases there reached a peak in 1992-93. Other countries were also affected. Currently, the number of new infections detected annually is low [119]. However, BSE remains a particular concern because cases still occur, albeit at a low rate, and there is a legacy arising from the prolonged incubation period of the disease, the life expectancy of cell banks, and the complexity of the processes by which they are established.

BSE in cattle has been transmitted to humans in the form of vCJD. Approximately 200 individuals have been affected either directly through exposure to BSE-infected material or through secondary transmission by non-leukocyte-depleted red blood cells. Classical CJD has also been transmitted by medical procedures, including administration of cadaveric growth hormone [120], corneal transplant, and the use of dura mater and vCJD may be transmissible by the same routes. vCJD has also been transmitted by human blood products. Although there is no evidence of vCJD transmission by plasma products, public health precautions have been implemented to minimize the possible risk of onward vCJD transmission by plasma products [121]. Cattle-derived proteins, including serum, have often been used in the growth of cells in culture and the production of biological products, including vaccines and recombinant products. Therefore, it is important to ensure that any ruminant-derived material used in biopharmaceutical manufacture is free of the agents that cause TSE. Moreover, as there is a possible but unquantifiable risk that cells can become infected by the agents of TSE, it is important that possibly contaminated ruminant material has been excluded from the start of the development of any cell line used. When there is insufficient traceability in the legacy of a cell line, a risk assessment should be undertaken to aid in decision-making about the suitability of the cell line for the intended use. There is currently no practical validated test that can be used for biological products or cell-line testing for the agents of TSE other than infection of susceptible species, where the experiments are very difficult because of the length of the incubation. More usable tests such as protein misfolding cyclic amplification (PMCA), which is analogous to PCR for nucleic acids, and epitope protection assays [121] are under investigation, but their performance characteristics when used to detect TSE agents in biological products or cell lines have not been defined. Strategies for minimizing risk have therefore focused so far on sourcing materials from countries believed to be at very low risk of infection and on substituting animal-derived materials with non-animal-derived materials.

B.11.4.1 *Infectivity categories of tissues*

Ruminant tissues are categorized by the World Health Organization and other scientific bodies such as the European Medicines Agency into three Categories (Category A - High infectivity, Category B - Lower infectivity, and Category C - no detectable infectivity) [55, 123]. Category A includes brain and Category C includes materials such as testes and bile. Assays of improved sensitivity have shown infectivity in tissues, such as muscle, previously thought to be free of infectious agents, and the implication is that while certain tissues contain large amounts of infectivity, many other tissues may contain low levels that are difficult to detect [55,123, 124, 125].

B.11.4.2 *Control measures, sourcing and traceability*

Where effective alternatives to ruminant-derived material are available, they should be used in cell culture/manufacturing procedures. Examples include: 1) cell culture medium free of animal material, 2) polysorbate and magnesium stearate of plant origin, 3) enzymes, such as rennet, of microbial origin (used in lactose production), and 4) recombinant insulin and synthetic amino acids. It should be noted, however, that recombinant materials may themselves be exposed to animal materials, so this potential should be considered when choosing recombinant materials as alternatives. However, it is not always possible to use ingredients free of animal materials, and raw materials of non-ruminant origin. For example, fetal bovine serum might have to be used in the development of cell lines or for fermentation. Under these circumstances, the raw materials should be sourced from countries classified by the World Organization for Animal Health (OIE) as negligible BSE risk or GBR I, as classified by the European Food Safety Authority (EFSA). Raw materials of Category C may be sourced from countries that are classified as controlled risk provided there is assurance that no cross-contamination with materials of Category A or B could have occurred during collection and processing, with the caveat that while they have undetectable levels of infectivity, it could conceivably be present. Manufacturers should maintain records, so that the finished product from any batch is traceable to the origin of any ruminant ingredient used in its manufacture that might pose a risk of exposure to a TSE agent, and each ruminant ingredient is traceable to the finished product. This includes ingredients used to develop and produce the MCB and WCB, and as far as is possible, to the derivation of the cell line itself. This traceability in both directions is important for appropriate regulatory action if new scientific research indicates that there is a TSE infectivity risk in the materials used, or if there is an association of the products' use with vCJD. Category A and B ruminant materials, originating from BSE-enzootic countries, should not be used in biologicals production under any circumstances. Because new BSE cases continue to occur despite feed bans, because suitable tests for TSE agents in raw-materials are not available, and because developments in scientific research indicate the presence of pathological prions in Category C materials, the best approach to TSE safety is not to use animal-derived protein. The next best approach is to source raw materials from countries classified as free of BSE, bearing in mind that cases may be detected in future.

B.11.4.3 *Tests*

Currently, there are no suitable screening tests available for TSE agents in raw materials of human/ruminant origin similar to serological or PCR assays for the screening for viral agents. Newer tests are being developed to screen for the presence of TSE agents in blood (such as PMCA, epitope-protection assay, *etc.*). Such tests, once validated, could eventually become suitable for the screening of raw materials and cell banks.

Approximately 15% of human TSEs are associated with inherited mutations in the PrP gene. These familial, but transmissible, TSEs are associated with around 30 known pathogenic mutations or with insertions and deletions in the octapeptide-repeat region of PrP [126]. The PrP gene of new human cell substrates should be sequenced to exclude the presence of these genetic changes.

B.12 Summary of tests for the evaluation and characterization of animal-cell substrates

The following sections provide an overview of tests that are recommended for the evaluation and characterization of animal-cell substrates proposed for use in the production of biological products. Not all of the tests are appropriate for all animal-cell substrates, but each of them should be considered and a determination made as to its applicability for a given cell substrate in the context of its use to manufacture a specific product. In addition, the point(s) at which a test should be applied needs to be rationalized. The overall testing strategy should provide assurance that risks have been mitigated to reasonable levels for the product and its intended use. The testing strategy should be agreed with the NRA/NCL.

B.12.1 *Cell Seed*

The cell seed generally is derived from a cell or tissue source of interest because of its potential utility in the development of a biological product. In some cases, the cell may be expected to serve as the substrate for the production of multiple products. Reference Cell Banks (see also A.5.3) would be considered cell seeds. The cell seed is usually of limited quantity so that extensive testing is not feasible. Some of the seed is therefore used to produce a supply of cells in a quantity that allows more extensive testing as well as providing a long-term source of cells for use in manufacturing. This secondary cell source is usually termed the master cell bank (MCB). However, the cell seed also may be used to produce additional low-passage material that can be banked (pre-MCB) and used to generate MCBs that then are characterized as described in this document.

Tests on the cell seed can be done at any point before the establishment of the MCB. Usually, those tests are limited to information essential to make the decision to commit resources to the preparation of a MCB. Such tests typically include: viability, morphology, identity (*e.g.*, karyotype, isoenzymes), and sterility (*e.g.*, bacterial, fungal, mycoplasma). These data serve as important background information, but they cannot substitute for the full characterization of the MCB.

B.12.2 *Master Cell Bank (MCB) and Working Cell Bank (WCB)*

The MCB generally will be developed to generate a sufficient quantity of cells to supply enough vials of cells to produce many WCBs over an extended period (usually years).

MCBs typically contain at least 200 vials and often 1000 or more. There also should be a sufficient number of vials in the WCB to provide material for the characterization of the cell line.

Some tests on the WCB are conducted on cells recovered directly from the bank itself; but other tests will be conducted on cells that have been propagated to a passage at or beyond the level that will be used for production. In addition, some tests may be appropriate to use as in-process control (IPC) tests. In such cases, they should be identified and described in the recommendations applicable to specific products.

Authors

The scientific basis for the revision of the Requirements published in WHO TRS 878 was developed at the meetings of the WHO Study Group on cell substrates in 2006 and 2007 attended by the following people:

Dr Peter Christian, NIBSC, Potters Bar, UK; Dr M. Deschamps, GSK, Rixensart, Belgium; Dr Rajeev M. Dhere, Serum Institute of India, Pune, India; Dr Clarice Hutchens, Pfizer, St. Louis, Missouri, USA; Dr José Lebron, Merck Research Labs., West Point, USA; Dr Ivana Knezevic, Quality and Safety and Standards, IVB/FCH/WHO, Dr Scott Lambert, Quality and Safety and Standards, IVB/FCH/WHO; Dr Andrew Lewis, CBER, Bethesda, USA; Dr Laurent Mallet, Sanofi Pasteur, Marcy L'Etoile, France; Dr P. Nandapalan, TGA, Woden, Australia; Dr Keith Peden, CBER, Bethesda, USA; Dr John Petricciani, Consultant, Palm Springs, USA; Dr Rebecca Sheets, NIH/NIAID, Bethesda, USA; Dr Jinho Shin, Quality and Safety and Standards, IVB/FCH/WHO; Dr Yeowon Sohn, KFDA, Seoul, Republic of Korea; Dr Glyn Stacey, NIBSC, Potters Bar, UK; Mrs C.A.M. van der Velden, Consultant, Groenekan, The Netherlands; Dr Ralph Wagner, Paul-Ehrlich Institute, Langen, Germany; Dr Omala Wimalaratne, Medical Research Institute, Colombo, Sri Lanka; and Dr David Wood, Coordinator, Quality and Safety and Standards, IVB/FCH/WHO.

Since then, several draft recommendations were prepared by the drafting group Dr John Petricciani, Consultant, Palm Springs, USA; Dr Rebecca Sheets, NIH/NIAID, Bethesda, USA; Dr Glyn Stacey, NIBSC, Potters Bar, UK; and Dr Ivana Knezevic, Quality and Safety and Standards, IVB/ FCH/ WHO, and were reviewed by the WHO Study Group on cell substrates in 2008.

Following the meeting of the Study Group in April 2009, Bethesda, USA, draft recommendations were revised taking into account information on the current manufacturing and regulatory practice provided at the meeting attended by the following participants:

Dr K.S. Ahn, KFDA, Seoul, Republic of Korea; Dr J.H. Blusch, Novartis, Basel, Switzerland; Dr da Silva Guedes Jr., Bio-Manguinhos/Fiocruz, Rio de Janeiro, Brazil; Dr Michel Deschamps, GSK, Wavre, Belgiuim; Dr Guanmu Dong, NICPBP, Beijing, People's Republic of China; Dr B. Gauvin, Amgen Inc., Thousand Oaks, USA; Dr H. Kavermann, Roch Diagnostics GmbH, Penzeberg, Germany; Dr Katherine King, FDA, Bethesda, USA; Dr Ivana Knezevic, Quality and Safety and Standards, IVB/ FCH/ WHO; Dr Andrew Lewis, CBER, Bethesda, USA; Dr Laurent Mallet, Sanofi Pasteur, Marcy L'Etoile, France; Dr Philip Minor, NIBSC, Potters Bar, UK; Dr P. Nandapalan, TGA, Woden, Australia; Dr David Onions, Invitrogen Corporation, Carlsbad, USA; Dr Keith Peden, CBER, Bethesda, USA; Dr John Petricciani, Consultant, Palm Springs,

USA; Ms E. Ika Prawahju, NADFC, Jakarta Pusat, Indonesia; Dr Rebecca Sheets, NIH/NIAID, Bethesda, USA; Dr Glyn Stacey, NIBSC, Potters Bar, UK; Dr Ralph Wagner; Paul-Ehrlich Institute, Langen, Germany; and Dr David Wood, Coordinator, Quality and Safety and Standards, IVB/FCH/WHO.

Based on the comments received from a broad range of regulators, manufacturers of vaccines and other biologicals and other relevant experts in 2009, draft recommendations were updated by the drafting group and posted on the WHO biologicals web site for public consultation from 4 to 31 May 2010.

The WHO/BS/10.2132 document was prepared by the drafting group at its meeting held on 1-3 June 2010 at WHO, Geneva, taking into account comments received from the reviewers as well as from the following meeting participants:

Dr Christoph Conrad, Quality and Safety and Standards, IVB/ FCH/ WHO; Dr HyeNa Kang, Quality and Safety and Standards, IVB/ FCH/ WHO; Dr Ivana Knezevic, Quality and Safety and Standards, IVB/ FCH/ WHO; Dr John Petricciani, Consultant, Palm Springs, USA; Dr Rebecca Sheets, NIH/NIAID, Bethesda, USA; Dr Philip Minor, NIBSC, Potters Bar, UK; Dr David Onions, Invitrogen Corporation, Carlsbad, USA; Dr Keith Peden, CBER, Bethesda, USA; Dr JinHo Shin, Quality and Safety and Standards, IVB/ FCH/ WHO; and Dr Glyn Stacey, NIBSC, Potters Bar, UK.

Further changes were made to WHO/BS/10.2132 by the Expert Committee on Biological Standardization, resulting in the present document.

Acknowledgements

Acknowledgements are due to the following experts for their useful comments during several rounds of the draft recommendations:

Dr. Rajeev M. Dhere, Mr. Subhash G. Bankar, and Mr. Vivek B. Vaidya, Serum Institute of India, India; Dr. Guanmu Dong, National Institute for the Control of Pharmaceutical and Biological Products (NICPBP) China; Elizabeth Ika Prawahju, National Agency of Drug and Food (NADFC), Indonesia; Dr Nagendram (Palan) Nandapalan BVSc, Therapeutic Goods Administration (TGA), Austria; International Federation of Pharmaceutical Manufacturers and Associations (Gay Gauvin, Amgen; Ray Field, Joe Kutza, Jonathan Liu, and David Lindsay, AstraZeneca / MedImmune; Kimberly Duffy, Paul Duncan, Lisa Plitnick, Lisette Vromans, and Jayanthi Wolf, Merck; Jürgen Hubert Blusch, Novartis; Ronald W Fedechko, and Stefanie Pluschkell, Pfizer; Holger Kavermann, Roche; Lorenz Scheppler, Crucell; Frédéric Mortiaux, GSK Biologicals; Benedicte Mouterde, and Mallet Laurent, Sanofi Pasteur; Ryoko Krause, IFPMA); Jim Robertson, NIBSC, UK; Parenteral Drug Association (Anthony Cundell, Schering-Plough; Jens-Peter Gregersen, Novartis; Leonard Hayflick, University of California San Francisco; Linda Hendrick, Centocor; Arifa Khan, FDA Center for Biologics Evaluation and Research; Kathryn King, FDA Center for Drug Evaluation and Research (Committee Chair); Robert Kozak, Bayer; Zhong Liu, Schering-Plough; Barbara Potts, Consultant; Michael Ruffing, Boehringer-Ingelheim; Sally Seaver, Seaver Associates LLC; Glyn Stacey, National Institute for Biological Standards and Control; Dominick Vacante, Centocor; Hannelore Willkommen, Regulatory Affairs and Biological Safety Consulting; Martin Wisher, BioReliance; Ruth Wolff, Biologics Consulting Group); Koen Brusselmans, and Geneviève Waeterloos, Scientific Institute of Public Health, Belgium, Jens-Peter Gregersen, Novartis Vaccines and Diagnostics, Germany; Barbara Potts, Nereus Pharmaceuticals, USA.

References

1. Requirements for the use of animal cells as *in vitro* substrates for the production of biologicals.In: WHO Expert Committee on Biological Standardization, 47th Report, World Health Organization, 1998. Annex 1. (WHO Technical Report Series, No. 878).
2. Hilleman, MR (1968). Cells, vaccines, and the pursuit of precedent. Nat Cancer Inst Mono 29:463-470.
3. Hayflick L, Plotkin S, Stevenson R. History of the acceptance of human diploid cell strains as substrates for human virus vaccine production. Developments in biological standardization, 1987,68:9-17.
4. (1959) Requirements for Poliomyelitis Vaccine (Inactivated). In Requirements for Biological Substances: 1. General Requirements for Manufacturing Establishments and Control Laboratories; 2. Requirements for Poliomyelitis Vaccine (Inactivated). Report of a Study Group. In WHO Technical Report Series No. 178 World Health Organization, Geneva.
5. Requirements for Poliomyelitis Vaccine (Inactivated). In Requirements for Biological Substances. Manufacturing and Control Laboratories. Report of a WHO Expert Group. In WHO Technical Report Series No. 323, 1966, World Health Organization, Geneva.
6. L Hayflick. History of the development of human diploid cell strains () in Proceedings, symposium on the characterization and uses of human diploid cell strains, 1963, 37- 54, Opatija
7. Hayflick, L. (2001) A brief history of cell substrates used for the preparation of human biologicals. Dev Biol 106, 5-23.
8. Proc Symp on Human Diploid Cell. Suggested methods for the management and testing of diploid cell culture used for virus vaccine production. Yugoslav Academy of Sciences and Art, Zagreb. P 213-216, 1970.
9. (1972) Requirements for Poliomyelitis Vaccine (Oral). In Requirements for Biological Substances No.7. In WHO Technical Report Series No. 486, World Health Organization, Geneva.
10. International Conference on Harmonization, Q5D, Derivation and Characterization of Cell Substrates Used for Production of Biotechnological/Biological Products, 1997, http://www.ich.org/LOB/media/MEDIA429.pdf.
11. CBER Guidance for Industry, Characterization and Qualification of Cell Substrates and Other Biological Starting Materials Used in the Production of Viral Vaccines for the Pervention of Infectious Diseases, 2010. http://www.fda.gov/downloads/BiologicsBloodVaccines/GuidanceComplianceRegulatoryInformation/Guidances/Vaccines/UCM202439.pdf
12. Report of a WHO Study Group on Biologicals, Acceptability of cell substrates for production of biologicals. WHO Technical Report Series 747, 1987.
13. Requirements for continuous cell lines used for biologicals production. In: WHO Expert Committee on Biological Standardization. Thirty-sixth Report. Geneva, World Health Organization, 1987, Annex 3 (WHO Technical Report Series, No. 745)

14. International Conference on Harmonization, Q5A, Viral Safety Evaluation of Biotechnology Products Derived from Cell Lines of Human or Animal Origin, http://www.ich.org/LOB/media/MEDIA425.pdf
15. International Conference on Harmonization, Q5B, Quality of Biotechnological Products: Analysis of the Expression Construct in Cells Used for Production of r-DNA Derived Protein Products, http://www.ich.org/LOB/media/MEDIA426.pdf
16. Schaeffer WI. Terminology associated with cell, tissue, and organ culture, molecular biology, and molecular genetics. Tissue Culture Association Terminology Committee. In Vitro Cell Dev Biol. 1990 Jan;26(1):97-101.
17. Laboratory animals in vaccine production and control: replacement, reduction and refinement. CFM Hendriksen Ed., Kluwer Academic Publishers, 1988, Dordrecht.
18. Jacobs, J.P. (1976) Updated results on the karyology of the WI-38, MRC-5 and MRC-9 cell strains. Dev Biol Stand 37, 155-156.
19. Jacobs, J.P., Magrath, D.I., Garrett, A.J., Schild, G.C. (1981) Guidelines for the acceptability, management and testing of serially propagated human diploid cells for the production of live virus vaccines for use in man. J Biol Stand 9, 331-342].
20. Petricciani, JC, Huang, CC, BA Rubin, DP Yang, LC Minecci, Z Kadanka, EM Earley, Karyology standards for rhesus diploid cell line DBS-FRhL-2. J Biol Stand, 1976, 43-49.
21. Schollmayer, E., Schafer, D., Frisch, B., Schleiermacher, E. (1981) High resolution analysis and differential condensation in RBA-banded human chromosomes. Hum Genet 59, 187-193.
22. Rønne, M. (1989) Chromosome preparation and high resolution banding techniques. A review. J Dairy Sci 72, 1363-1377.
23. Healy LE, Ludwig TE, Choo A. International banking: checks, deposits, and withdrawals. Cell Stem Cell. 2008 Apr 10;2(4):305-6.
24. Laflamme MA, Chen KY, Naumova AV, Muskheli V, Fugate JA, Dupras SK, Reinecke H, Xu C, Hassanipour M, Police S, O'Sullivan C, Collins L, Chen Y, Minami E, Gill EA, Ueno S, Yuan C, Gold J, Murry CE. Cardiomyocytes derived from human embryonic stem cells in pro-survival factors enhance function of infarcted rat hearts. Nat Biotechnol. 2007 Sep;25(9):1015-24. Epub 2007 Aug 26.
25. Fleckenstein B, Daniel MD, Hunt RD. Tumor inductions with DNA of oncogenic primate herpesviruses. Nature, 1978, 274:57-9.
26. Petricciani JC, Regan PJ. Risk of neoplastic transformation from cellular DNA: calculations using the oncogene model. Developments in biological standardization, 1986, 68:43-49.
27. Temin HM. Overview of biological effects of addition of DNA molecules to cells. Journal of Medical Virology, 1990, 31: 13-17.
28. Wierenga DE, Cogan J, Petriccaini JC. Administration of tumor cell chromatin to immunosuppressed and non-immunosuppressed primates. Biologicals, 1995,23:221-24.
29. Petricciani JC, Horaud FN. DNA, dragons, and sanity. Biologicals, 1995, 23:233-238.
30. Nichols WW et al, Potential DNA vaccine integration into host cell genome. Annals of the New York Academy of Science, 1995, 772:30-38.

31. Kurth R. Risk potential of the chromosomal insertion of foreign DNA. Annals of the New York Academy of Sciences, 1995,772:140-150.
32. Coffin JM. Molecular mechanisms of nucleic acid integration. Journal of medical virology, 1990, 31:43-49.
33. Sheng L, Cai, F, Zhu, Y, Pal, A, Athanasiou, M, Orrison, B, Blair, DG, Hughes, SH, Coffin, JM, Lewis, AM, Peden, K (2008) Oncogenicity of DNA *in vivo*: Tumor induction with expression plasmids for activated H-ras and c-myc. Biologicals 36, 184-197.
34. Garnick, R.L. (1996) Experience with viral contamination in cell culture. Dev Biol Stand 88, 49-56.
35. Contamination of the CHO cells by orbiviruses: Burstyn DG, Dev Biol Stand. 1996;88:199-203.
36. Contamination of genetically engineered Chinese hamster ovary cells.: Rabenau H, Ohlinger V, Anderson J, Selb B, Cinatl J, Wolf W, Frost J, Mellor P, Doerr HW., Biologicals. 1993 Sep;21(3):207-14.
37. Sheng-Fowler L, Lewis AM Jr, Peden K. Quantitative determination of the infectivity of the proviral DNA of a retrovirus *in vitro*: Evaluation of methods for DNA inactivation. Biologicals. 2009 Aug;37(4):259-69.
38. Lewis, AM, Jr., Krause, P, Peden, K (2001), A defined-risks approach to the regulatory assessment of the use of tumourigenic cells as substrates for viral vaccine manufacture. Dev Biol 106, 513-535.
39. Krause, P.R., Lewis, A.M., Jr. (1998) Safety of viral DNA in biological products. Biologicals 26, 317-320.
40. Li Sheng-Fowler, Andrew M. Lewis Jr., Keith Peden. Issues associated with residual cell-substrate DNA in viral vaccines. Biologicals 37 (2009) 190-195
41. Ledwith, B.J., Manam, S., Troilo, P.J., Barnum, A.B., Pauley, C.J., Griffiths, I.T., Harper, L.B., Beare, C.M., Bagdon, W.J., Nichols, W.W. (2000) Plasmid DNA Vaccines: Investigation of Integration into Host Cellular DNA following Intramuscular Injection in Mice. Intervirology 43, 258-272.
42. Li Sheng-Fowler, Fang Cai, Haiqing Fu, Yong Zhu, Brian Orrison1, Gideon Foseh, Don G. Blair, Stephen H. Hughes, John M. Coffin, Andrew M. Lewis Jr and Keith Peden. *Int. J. Biol. Sci.* 2010; 6(2):151-162
43. Burns, P.A., Jack, A., Neilson, F., Haddow, S., Balmain, A. (1991). Transformation of mouse skin endothelial cells *in vivo* by direct application of plasmid DNA encoding the human T24 H-ras oncogene. Oncogene 6, 1973-1978.
44. Meng, F., Henson, R., Wehbe-Janek, H., Ghoshal, K., Jacob, S.T., Patel, T. (2007) MicroRNA-21 regulates expression of the PTEN tumor suppressor gene in human hepatocellular cancer. Gastroenterology 133, 647-658.
45. Ma, L., Teruya-Feldstein, J., Weinberg, R.A. (2007) Tumour invasion and metastasis initiated by microRNA-10b in breast cancer. Nature 449, 682-688.
46. He, L., Thomson, J.M., Hemann, M.T., Hernando-Monge, E., Mu, D., Goodson, S., Powers, S., Cordon-Cardo, C., Lowe, S.W., Hannon, G.J., Hammond, S.M. (2005) A microRNA polycistron as a potential human oncogene. Nature 435, 828-833.
47. Hayashita, Y., Osada, H., Tatematsu, Y., Yamada, H., Yanagisawa, K., Tomida, S., Yatabe, Y., Kawahara, K., Sekido, Y., Takahashi, T. (2005) A polycistronic

microRNA cluster, miR-17-92, is overexpressed in human lung cancers and enhances cell proliferation. Cancer Res 65, 9628-9632.
48. Israel, M.A., Chan, H.W., Hourihan, S.L., Rowe, W.P., Martin, M.A. (1979) Biological activity of polyoma viral DNA in mice and hamsters. J Virol 29, 990-996.
49. Peden, K., Sheng, L., Pal, A., Lewis, A. (2006) Biological activity of residual cell-substrate DNA. Dev Biol (Basel) 123, 45-53.
50. Perrin, P., Morgeaux, S. (1995) Inactivation of DNA by beta-propiolactone. Biologicals 23, 207-211.
51. Lebron, J.A., Troilol, P.J., Pacchione, S., Griffiths, T.G., Harper, L.B., Mixson, L.A., Jackson, B.E., Michna, L., Barnum, A.B., Denisova, L., Johnson, C.N., Maurer, K.L., Morgan-Hoffman, S., Niu, Z., Roden, D.F., Wang, Z., Wolf, J.J., Hamilton, T.R., Laux, K.M., Soper, K.A., Ledwith, B.J. (2006) Adaptation of the WHO guideline for residual DNA in parenteral vaccines produced on continuous cell lines to a limit for oral vaccines. Dev Biol (Basel) 123, 35-44.
52. Coecke, S., Balls, M., Bowe, G., Davis, J., Gstraunthaler, G., Hartung, T., Hay, R., Merten, O-W., Price, A., Shechtman, L., Stacey, G.N. and Stokes, W., (2005), Guidance on Good Cell Culture Practice. A report of the second ECVAM Task Force on Good Cell Culture Practice, ATLA, **33**: 1-27
53. Consensus Guidance for Banking and Supply of Human Embryonic Stem Cell Lines for Research Purposes, J Stem Cell Reviews and Reports, 2009 5 301-314.
54. Stacey (2007) Risk Assessment of cell Culture processes. Medicines from Animal Cells, Eds Stacey GN and Davis JM, J Wiley & sons, Chichester, 27 April 2007, ISBN 9780470850947
55. WHO guidelines on tissue infectivity distribution in transmissible spongiform encephalopathies. WHO 2006. http://www.who.int/bloodproducts/tse/WHO%20TSE%20Guidelines%20FINAL-22%20JuneupdatedNL.pdf
56. Plavsic ZM, Bolin S. Resistance of porcine circovirus to gamma irradiation. BioPharm, 2001.
57. Requirements for the collection, processing and quality control of blood, blood components and plasma derivatives. In: WHO Expert Committee on Biological Standardization. Forty-third report. Geneva, World Health Organization, 1994, Annex 2 (WHO Technical Report Series, No. 840).
58. G N Stacey, J R Masters, Cryopreservation and banking of mammalian cell lines. Nature protocols 2008;3(12):1981-9.
59. Chu L, Robinson DK. Industrial choices for protein production by large-scale cell culture. Current Opinion in Biotechnology 2001, 12:180-187.
60. Ho Y, Varley J, Mantalaris A. Development and analysis of a mathematical model for Antibody-producing GS-NSO cells under normal and hyperosmotic culture conditions. Biotechnol Progr 2006, 22, 1560-1569.
61. McPherson I, Montagnier I. Agar suspension culture for the selective assay of cells transformed by polyoma virus. Virology 23:291-294;1964.
62. Petricciani JC, Levenbook I, Locke, R. Human muscle: a model for the study of human neoplasia. Investigational New Drugs 1:297-302;1983.
63. Furesz J, Farok A, Contreras G, Becker B. Tumorigencity testing of various cell substrates for production of biologicals. Develop Biol Stand 70:233-243;1989.

64. Manohar M, Orrisoam B, Peden K, Lewis AM. Assessing the tumourigenic phenotype of Vero cells in adult and newborn nude mice. Biologicals 36:65-72;2008.
65. Giovanella BC, Yim SO, Stehlin JC, Williams LJ. Development of invasive tumors in nude mice after injection of cultured human melanoma cell. J Natl Cancer Inst 48:15311-1534;1972.
66. Manu Manohar1, Brian Orrison, Keith Peden, Andrew M. Lewis Jr. Assessing the tumourigenic phenotype of VERO cells in adult and newborn nude mice. Biologicals 36 (2008) 65-72
67. Hentze H, Soong PL, Wang ST, Phillips BW, Putti TC, Dunn NR. Teratoma formation by human embryonic stem cells: Evaluation of essential parameters for future safety studies. Stem Cell Res. 2009 Feb 12.
68. van Steenis G, van Wezel AL. Use of ATG-treated newborn rat for *in vivo* tumourigenicity testing of cell substrates. Dev Biol Stand 50:37-46;1982.
69. Noguchi PD, Johnson JB, Petricciani JC. Comparison of the nude mouse and immunosuppressed newborn hamsters a quantitative tumourigenicity test for human cell lines. IRCS Medical Science: biomedical tech: cancer: cell and membrane biology: experimental animals: immunology and allergy: Pathology, 7:302-4; 1979.
70. Wallace R, Vasington PJ, Petricciani JC. Heterotransplantation of cultured cell lines in newborn hamsters treated with antilymphocyte serum. Nature 230:454-455;1971.
71. Foley GE, Handler AH, Adams RA, Craig, JM. Assessment of potential malignancy of cultured cells: further observations on the differentiation of Normal" and "Neoplastic" cells maintained *in vitro* by heterotransplantation in Syrian hamsters. Natl Cancer Inst Monograph 7:173-204;1962.
72. Petricciani JC, Martin DP. Use of nonhuman primates for assaying tumourigenicity of viral vaccine cell substrates. Transplant Proc 6:189-192;1974.
73. Levenbook I, Petricciani JC, Qi Y, Elisberg BL, Rogers L, Jackson LB, Wierenga DE, Webster BA. Tumourigenicity testing of primate cell lines in nude mice, muscle organ culture and for colony formation in soft agarose. J Biol Standard 13:135-141;1985.
74. Levenbook I, Smith PL, Petricciani JC. A comparison of three routes of inoculation for testing tumourigenicty of cell lines in nude mice. J Biol Stand 9:75-80, 1981.
75. Andrew M. Lewis Jr, David W. Alling, Steven M. Banks, Silvia Soddu, James L. Cook. Evaluating virus-transformed cell tumourigenicity. Journal of Virological Methods 79 (1999) 41–50
76. Pulciani, S., Santos, E., Lauver, A.V., Long, L.K., Barbacid, M. (1982) Transforming genes in human tumors. J Cell Biochem 20, 51-61.
77. Murray, M.J., Shilo, B.Z., Shih, C., Cowing, D., Hsu, H.W., Weinberg, R.A. (1981) Three different human tumor cell lines contain different oncogenes. Cell 25, 355-361.
78. Shih, C., Padhy, L.C., Murray, M., Weinberg, R.A. (1981) Transforming genes of carcinomas and neuroblastomas introduced into mouse fibroblasts. Nature 290, 261-264.
79. Wood DJ, Minor PD. (1990) Meeting report: Use of human diploid cells in vaccine production. Biologicals 18:143-146.
80. An international system for human cytogenetic nomenclature (2005) Shaffer LG and Tommerup N. Karger, editors, 2005.

81. Annex 1. Recommendations for the production and control of poliomyelitis vaccine (oral)(Addendum 2000). In World Health Organization Technical Report Series, Volume 910, World Health Organization, Geneva.
82. (1994) Requirements for measles, mumps and rubella vaccines and combined vaccine (Live) (Requirements for Biological Substances No. 47. 1992) In: WHO Expert Committee on Biological Standardization. Forty-fourth Report. In WHO Technical Report Series.
83. Nicklas W., Kraft V., Meyer, B. (1993) Contamination of transplantable tumors, cell lines, and monoclonal antibodies with rodent viruses. Laboratory Animal Science 43 (4), 296-300.
84. Collins, M.J., Parker, J. C. (1972) Murine virus contaminants of leukemia viruses and transplantable tumors. Journal of the National Cancer Institute 49 (4), 1139-1143.
85. Bierley, S.T., Raineri, R., Poiley, J.A., Morgan, E.M. (1996) A comparison of methods for the estimation of retroviral burden. Dev Biol Stand 88, 163-165.
86. Goff SP, Chapter 55: Retroviridae: The Retroviruses and their Replication, Fields Virology, 5th edition edited by Knipe DM, Howley PM, Griffin DE, Martin MA, Lamb RA, Lippincott, Williams, & Wilkins, Baltimore, MD, 2007, p. 2001
87. Arnold, B.A., Hepler, R.W., Keller, P.M. (1998) One-step fluorescent probe product-enhanced reverse transcriptase assay. Biotechniques 25, 98-106.
88. Maudru, T., Peden, K.W. (1998) Adaptation of the fluorogenic 5'-nuclease chemistry to a PCR-based reverse transcriptase assay. Biotechniques 25, 972-975.
89. Lovatt, A., Black, J., Galbraith, D., Doherty, I., Moran, M.W., Shepherd, A.J., Griffen, A., Bailey, A., Wilson, N., Smith, K.T. (1999) High throughput detection of retrovirus-associated reverse transcriptase using an improved fluorescent product enhanced reverse transcriptase assay and its comparison to conventional detection methods. J Virol Methods 82, 185-200.
90. Sears, J.F., Khan, A.S. (2003) Single-tube fluorescent product-enhanced reverse transcriptase assay with Ampliwax (STF-PERT) for retrovirus quantitation. J Virol Methods 108, 139-142.
91. Pyra, H., Böni, J., Schüpbach, J. (1994) Ultrasensitive retrovirus detection by a reverse transcriptase assay based on product enhancement. Proc Natl Acad Sci U S A 91, 1544-1548.
92. Böni, J., Pyra, H., Schüpbach, J. (1996) Sensitive detection and quantification of particle-associated reverse transcriptase in plasma of HIV-1-infected individuals by the product- enhanced reverse transcriptase (PERT) assay. J Med Virol 49, 23-28.
93. Böni, J., Stalder, J., Reigel, F., Schüpbach, J. (1996) Detection of reverse transcriptase activity in live attenuated virus vaccines. Clin Diagn Virol 5, 43-53.
94. Weissmahr, R.N., Schüpbach, J., Böni, J. (1997) Reverse transcriptase activity in chicken embryo fibroblast culture supernatants is associated with particles containing endogenous avian retrovirus EAV-0 RNA. J Virol 71, 3005-3012.
95. Khan, A.S., Maudru, T., Thompson, A., Muller, J., Sears, J.F., Peden, K.W.C. (1998) The reverse transcriptase activity in cell-free medium of chicken embryo fibroblast cultures is not associated with a replication-competent retrovirus. J Clin Virol 11, 7-18.

96. Maudru, T., Peden, K. (1997) Elimination of background signals in a modified polymerase chain reaction-based reverse transcriptase assay. J Virol Methods 66, 247-261.
97. Lugert, R., König, H., Kurth, R., Tönjes, R.R. (1996) Specific suppression of false-positive signals in the product-enhanced reverse transcriptase assay. Biotechniques 20, 210-217.
98. Voisset, C., Tönjes, R.R., Breyton, P., Mandrand, B., Paranhos-Baccalà, G. (2001) Specific detection of RT activity in culture supernantants of retrovirus-producing cells, using synthetic DNA as competitor in polymerase enhanced reverse transcriptase assay. J Virol Methods 94, 187-193.
99. Fan, X.Y., Lu, G.Z., Wu, L.N., Chen, J.H., Xu, W.Q., Zhao, C.N., Guo, S.Q. (2006) A modified single-tube one-step product-enhanced reverse transcriptase (mSTOS-PERT) assay with heparin as DNA polymerase inhibitor for specific detection of RTase activity. J Clin Virol 37, 305-312.
100. Peebles, P.T. (1975) An *in vitro* focus-induction assay for xenotropic murine leukemia virus, feline leukemia virus C, and the feline--primate viruses RD-114/CCC/M-7. Virology 67, 288-291.
101. Sommerfelt, M.A., Weiss, R.A. (1990) Receptor interference groups of 20 retroviruses plating on human cells. Virology 176, 58-69.
102. Onions D, Kolman J Massively parallel sequencing, a new method for detecting adventitious agents. Biologicals. 2010 May;38(3):377-80. Epub 2010 Mar 24
103. Wang D, Urisman A, Liu YT, Springer M, Ksiazek TG, Erdman DD, Mardis ER, Hickenbotham M, Magrini V, Eldred J, Latreille JP, Wilson RK, Ganem D, DeRisi JL. PLoS Biol. 2003 Nov;1(2):E2. Epub 2003 Nov 17. Viral discovery and sequence recovery using DNA microarrays.
104. Ecker, D.J., Sampath, R., Massire, C., Blyn, L.B., Hall, T.A., Eshoo, M.W., Hofstadler, S.A. (2008) Ibis T5000: a universal biosensor approach for microbiology. Nat Rev Microbiol 6, 553-558.
105. Allander T, Tammi MT, Eriksson M, Bjerkner A, Tiveljung-Lindell A, Andersson B. Cloning of a human parvovirus by molecular screening of respiratory tract samples. Proc Natl Acad Sci U S A 2005;102(43):15712.
106. Gaynor AM, Nissen MD, Whiley DM, Mackay IM, Lambert SB, Wu G, et al. Identification of a novel polyomavirus from patients with acute respiratory tract infections. PLoS Pathog 2007;3(5):e64.
107. Allander T, Emerson SU, Engle RE, Purcell RH, Bukh J. A virus discovery method incorporating DNase treatment and its application to the identification of two bovine parvovirus species. Proc Natl Acad Sci U S A 2001;98(20):11609–14.
108. Allander T, Andreasson K, Gupta S, Bjerkner A, Bogdanovic G, Persson MA, et al. Identification of a third human polyomavirus. J Virol 2007;81(8): 4130–6.
109. Victoria JG, Wang C, Jones MS, Jaing C, McLoughlin K, Gardner S, Delwart EL. Viral nucleic acids in live-attenuated vaccines: detection of minority variants and an adventitious virus. J Virol. 2010 Jun;84(12):6033-40. Epub 2010 Apr 7.
110. Sampath, R., Russell, K.L., Massire, C., Eshoo, M.W., Harpin, V., Blyn, L.B., Melton, R., Ivy, C., Pennella, T., Li, F., Levene, H., Hall, T.A., Libby, B., Fan, N., Walcott, D.J., Ranken, R., Pear, M., Schink, A., Gutierrez, J., Drader, J., Moore, D., Metzgar, D., Addington, L., Rothman, R., Gaydos, C.A., Yang, S., St George, K., Fuschino,

M.E., Dean, A.B., Stallknecht, D.E., Goekjian, G., Yingst, S., Monteville, M., Saad, M.D., Whitehouse, C.A., Baldwin, C., Rudnick, K.H., Hofstadler, S.A., Lemon, S.M., Ecker, D.J. (2007) Global surveillance of emerging Influenza virus genotypes by mass spectrometry. PLoS ONE 2, e489.

111. General Requirements for the Sterility of Biological Substances (Requirements for Biological Substances No. 6 revised 1973) In: WHO Expert Committee on Biological Standardization. Twenty-fifth Report. Annex 4. In WHO Technical Report Series No. 530 World Health Organization, Geneva.
112. Sterility test. 21CFR 610.12
113. 71. Sterility testing. United States Pharmacopeia
114. 2.6.1. Sterility. European Pharmacopoeia 6.0
115. General Requirements for the Sterility of Biological Substances (Requirements for Biological Substances No. 6 revised 1973) In: WHO Expert Committee on Biological Standardization. Forty-sixth Report. 1998 Annex 3. In WHO Technical Report Series No. 872 World Health Organization, Geneva.
116. European Pharmacopoeia 6th Edition, 2007.
117. Kovacs, G.G., Budka, H. (2008) Prion diseases: from protein to cell pathology. Am J Pathol 172, 555-565.
118. Ryou, C. (2007) Prions and prion diseases: fundamentals and mechanistic details. J Microbiol Biotechnol 17, 1059-1070.
119. Annual incidence rate of bovine spongiform encephalopathy (BSE) in OIE Member Countries that have reported cases, excluding the United Kingdom. World Organisation for Animal Health. http://www.oie.int/eng/info/en_esbincidence.htm.
120. Preece M.A. 1991 Creutzfeldt-Jakob disease following treatment with human pituitary hormones. Clinical Endocrinology 34: 527-529.
121. C. M. Millar, N. Connor, G. Dolan, C. A. Lee, M. Makris, J. Wilde, M. Winter, J. W. Ironside, N. Gill, and F. G. H. Hill. Risk reduction strategies for variant Creutzfeldt–Jakob disease transmission by UK plasma products and their impact on patients with inherited bleeding disorders. Haemophilia, 2010, 16(2): 305-15.
122. Minor P and Brown P (2006) Diagnostic Tests for Antemortem Screening of Creutzfeldt-Jakob Disease. 5: 119-148. In Creutzfeldt-Jakob Disease: Managing the Risk of Transmission by Blood, Plasma, and Tissues, Ed M.L Turner, AABB, Bethesda, MD.
123. WHO tables on guidelines on tissue infectivity distribution in transmissible spongiform encephalopathies. WHO 2010. http://www.who.int/bloodproducts/tablestissueinfectivity.pdf
124. Updated opinion on TSE infectivity distribution in ruminant tissues. European Commission, 2002. http://ec.europa.eu/food/fs/sc/ssc/out296_en.pdf.
125. Note for guidance on minimizing the risk of transmitting animal spongiform encephalopathy agents via human and veterinary medicinal products adopted by the Committee for proprietary medicinal products and by the Committee for veterinary medicinal products. Official Journal of the Duropean Union, 2004. http://www.ema.europa.eu/pdfs/human/bwp/TSE%20NFG%20410-rev2.pdf.
126. McKintosh E, Tabrizi SJ, Collinge J. Prion diseases. J Neurovirol. 2003 Apr;9(2):183-93.

Appendix 1

Tests for Bovine Viruses in Serum Used to Produce Cell Banks

Serum should be tested for adventitious agents, such as bacteria, fungi, mycoplasmas, and viruses, prior to use in the production of MCBs and WCBs. In addition, consideration should be given to risk-mitigation strategies, such as inactivation by heat or irradiation, to ensure that adventitious agents that were not detected in the manufacture and quality control of the serum would be inactivated to a degree acceptable to the NRA/NCL. If irradiation or other inactivation (*e.g.*, heat sterilization) methods are used in the manufacture of the serum, the tests for adventitious agents should be performed prior to inactivation to enhance the opportunity for detecting the contamination. If evidence of viral contamination is found in any of the tests, the serum is acceptable only if the virus is identified and shown to be present in an amount that has been shown in a validation study to be effectively inactivated. For serum that is not to be subjected to a virus inactivation / removal procedure, if evidence of viral contamination is found in any tests, generally, the serum would not be acceptable. If the manufacturer chooses to use serum that has not been inactivated, thorough testing of the serum for adventitious agents using current best practices should be undertaken. If any viruses are identified in the serum, the cell banks made in this manner should be shown to be free of the identified virus(es).

If irradiation is used, it is important to ensure that a reproducible dose is delivered to all batches and the component units of each batch. The irradiation dose must be low enough so that the biological properties of the reagents are retained while being high enough to reduce virological risk. Therefore, irradiation delivered at such a dose may not be a sterilizing dose.

Factors to be considered in testing serum

Bovine serum can be contaminated by a wide range of viruses. Manufacturers typically produce very large pools of serum involving samples from up to a thousand animals. Consequently, many serum batches contain detectable, genomic sequences of viruses like bovine viral diarrhoea virus (BVDV) and bovine polyomavirus (BPyV) [1] although, this might represent contamination of the pool by one viraemic animal.

Other viruses are sporadic contaminants and may be regionally restricted like Cache Valley virus, bluetongue virus and epizootic haemorrhagic disease virus. In some cases, contamination has only been reported on a few occasions, as in the case of calicivirus 2117 [2].

Application of new methods, like massively parallel sequencing, has revealed new viruses, like parvoviruses, some of which are frequent and high level contaminants of serum [3, 4]. The importance and potential pathogenicity of these viruses requires further investigation.

An important factor in infectivity assays is that virions might be neutralised by antibody in the serum pool. It is advisable to set limits for the level of BVDV neutralising antibody in serum pools as this may mask the presence of potentially infectious virus.

There should be an awareness of the statistical limits of screening assays to detect viruses in large serum pools. For example, in an infection of a fermenter by Cache Valley virus it was estimated that less than 10 viruses per litre were present in the serum and, at this low level, the virus escaped detection by conventional screening methods [5].

General screening assay for infectious viruses

A general screening assay typically involves culturing indicator cells over 21 days with test serum at 15% in the medium. At least 2 sub-passages of the cells should be undertaken, usually at days 7 and 14. Detection of virus infection involves regular examination for the development of a cytopathic effect, haemadsorption assays, and immunofluorescence (or other appropriate immunological detection method) for specific viruses. Immunofluorescence (IF) is particularly important for the detection of BVDV as many isolates are non-cytopathic. At the end of the assay, cytological staining (*e.g.,* with Giemsa stain) is used to reveal viral inclusions and other cytopathic effects not detected during the direct observation of the live cells.

Indicator cells should be selected that are permissive for a wide range of bovine viruses. Madin-Darby Bovine kidney Cells (MDBK) or, Bovine Turbinate cells, are often used and it is also of value to include additional cells like Vero cells.

The assay should be capable of detecting: bovine viral diarrhea virus (BVDV), bovine parainfluenza type 3 Virus (BPIV3), bovine parvovirus 1 (BPV), rabies virus (RV), reovirus 3 (REO3), infectious bovine rhinotracheitis Virus (IBR), bovine respiratory syncytial virus (BRSV), blue tongue virus (BTV), bovine adenovirus 5 (BAV5), and vesicular stomatitis virus (VSV). Separate, positive control bottles, of indicator cells should be infected with each of the viruses above, except rabies virus. In the case of rabies virus, slides of fixed, infected, cells should be used as a positive control for the immunofluorescent assay. Uninfected negative control cells should also be established.

A typical assay involves the use of 75cm^2 bottles containing the indicator cells and a total of ~250 ml of test serum, allowing for serum used during re-feeding the cells after passage.

Procedure

Assay set up

Initially, negative control bottles and test article bottles are established. The test article bottles are inoculated and maintained with the test serum at 15% in the medium. The negative control bottles are mock infected with serum known to be free of detectable viruses. Passage of the cells is usually required on day 7.

Cells for the positive control are prepared from the negative control bottles on day 13 or 14 or when the cells are ≥70% confluent. The cells are subcultured into 25 cm^2 flasks (for IF) and 6-well plates [for haemadsorption (HAd) and cytological staining (CS)].

The following day, the remaining negative control and test article cells are subcultured to 75 cm^2 flasks for IF and to 6-well plates for haemadsorption (HAd) and cytological staining (CS).

Infection with positive controls

Coincident with the final subculture of test article and negative control cultures, flasks of bovine turbinate cells (BT) are inoculated with the IF positive control viruses BVDV, BAV5, BPV, BTV, BRSV, IBR, and BPIV-3. Plates of BT cells are inoculated with BPIV-3, the positive control for haemadsorption, and with cytopathic BVDV, the positive control for cytological staining. Likewise, Vero flasks are inoculated with REO3, the IF positive control, and plates are inoculated with BPIV-3, the haemadsorption and cytological staining positive control. All IF positive control viruses should be inoculated at 100-300 $TCID_{50}$.

Analysis

After a minimum of 21 days post-inoculation, and at least 7 days after the last subculture, (but earlier if CPE is observed), negative control and test article cultures are assayed for HAd and fixed for IF and CS.

Cells from the positive control flasks are transferred to multi-well slides and fixed for IF when CPE involving ≥10% of the monolayer is observed and stored at ≤ -60°C. Cells in the positive control 6-well plates are assayed for HAd and CS 7 days after inoculation or when CPE is apparent. HAd involves testing at least one 6 well plate with chicken and guinea pig erythrocytes at 2-8°C and at 20-25°C.

NAAT assays for viruses

Nucleic acid amplification technologies like PCR have utility in screening serum for sporadic contaminants and for those viruses where infectivity assays are not available. Nucleic acid extractions should be from a significant volume (*e.g.*, 25 to 50ml) and the statistical limits for detection in the serum pool should be calculated. The presence of genomic sequences does not necessarily indicate the presence of infectious virus, although encapsidated genomes can be identified by treatment of the sample with nucleases prior to amplification. Some virus inactivating or removal processes can be evaluated using NAAT, by determining if intact, full-length, amplifiable, genomes are present before and after treatment.

Specific in vitro infectivity assays

Bovine polyomavirus is an important contaminant because it is able to infect primate cells [6], belongs to an oncogenic family of viruses and expresses a T-antigen that can transform primary cells into tumour cells [7]. Furthermore, there is serological evidence of zoonotic infection [8]. Infectious virus is not easily detected in conventional assays, a long period of culture and a NAAT endpoint or, immunological endpoint like immunofluorescence, should be used.

Other viruses are not easily detected in standard infectivity assays. For instance calicivirus 2117 appears to be more permissive for replication in CHO cells than standard bovine cell lines used in *in vitro* infectivity assays. Similarly, while general screening methods will detect certain bovine adenoviruses, herpesviruses and parvoviruses, not all bovine viruses belonging to these families are detected.

References

1. Schuurman R, van Steenis B, van Strien A, van der Noordaa J, Sol C. Frequent detection of bovine polyomavirus in commercial batches of calf serum by using the polymerase chain reaction. J Gen Virol. 1991 Nov;72 (Pt 11):2739-45.
2. Oehmig A, Büttner M, Weiland F, Werz W, Bergemann K, Pfaff E. Identification of a calicivirus isolate of unknown origin.J Gen Virol. 2003 Oct;84(Pt 10):2837-45.
3. Allander T, Emerson SU, Engle RE, Purcell RH, Bukh J. A virus discovery method incorporating DNase treatment and its application to the identification of two bovine parvovirus species. Proc Natl Acad Sci USA. 2001 Sep 25;98(20):11609-14. Epub 2001 Sep 18.
4. Onions D, Kolman J. Massively parallel sequencing, a new method for detecting adventitious agents. Biologicals. 2010 May;38(3):377-80. Epub 2010 Mar 24.
5. Nims RW, Dusing SK, Hsieh W-T, Lovatt A, Reid G, Onions D, Milne EW. Detection of Cache Valley Virus in Biologics Manufactured in CHO Cells. BioPharm International. Vol. 21, Issue 10 (2008).

6. Richmond JE, Parry JV, Gardner SD Characterisation of a polyomavirus in two foetal rhesus monkey kidney cell lines used for the growth of hepatitis A virus. Arch Virol. 1984;80(2-3):131-46.
7. Schuurman R, van Strien A, van Steenis B, van der Noordaa J, Sol C. Bovine polyomavirus, a cell-transforming virus with tumourigenic potential. J Gen Virol. 1992 Nov;73 (Pt 11):2871-8.
8. Parry JV, Gardner SD. Human exposure to bovine polyomavirus: a zoonosis? Arch Virol. 1986;87(3-4):287-96

Appendix 2

Tumourigenicity Protocol Using Athymic Nude Mice to Assess Mammalian Cells

During the characterization of a MCB (or WCB), the cells should be examined for tumourigenicity in a test approved by the national regulatory authority (NRA) or the national control authority (NCA).

The following model protocol is provided to assist manufacturers and NRAs/NCLs in standardizing the tumourigenicity testing procedure so that the interpretation and comparability of data among various laboratories and regulatory authorities can be facilitated.

1. **Test Animals**
   The test article cell line and the control cells are each injected into separate groups of 10 athymic mice (Nu/Nu genotype) 4-7 weeks old.

   Because male athymic mice often display aggressive traits against each other when housed together, loss of some mice during the observation period occurs often. Therefore, the use of only female mice should be considered.

2. **Test article cells**
   Cells from the MCB or MWCB that have been propagated to at least 3 population doublings beyond the limit for production are examined for tumourigenicity.

3. **Control cells**
   a. **Positive control cells**
      HeLa cells from the WHO cell bank are recommended as the positive control reference preparation. Portions of that bank are stored at the ATCC (USA) and NIBSC (UK).

      Other cells may be acceptable to the NRA/NCL if HeLa cells from the WHO cell bank are not available.

   b. **Negative control cells**
      Negative control cells are not required. Databases (both published data and the unpublished records/data of the animal production facility that supplied the test animals) of rates of spontaneous neoplastic diseases in nude mice may be taken into account during the assessment of the results of a tumourigenicity test.

If negative control cells are included, clear justification must be provided. In particular, the number of animals used must provide meaningful data, and the rationale for generating additional data must be persuasive to the NRA/NCL in the context of animal welfare regulations.

4. **Validity**
In a valid test, progressively growing tumors should be produced in at least 9 of 10 animals injected with the positive control reference cells. At least 90% of the inoculated control and cells and test cells must be viable for the test to be valid.

5. **Inoculum**
The inoculum for each animal is 10^7 viable cells (except as described in 11.b), suspended in a volume of 0.1 mL of PBS.

Cell culture media without serum has been used in the past to suspend the cell inoculum. However, many current media are serum-free and contain one or more growth factors that may affect the result of the tumourigenicity assay. Therefore careful consideration should be given to the choice of the liquid into which the cells are suspended.

6. **Injection route and site**
The injection of cells may be by either the intramuscular (IM) or subcutaneous (SC) route. If the IM route is selected, the cells should be injected into the thigh of one leg. If the SC route is selected, the cells should be injected into the supraclavicular region of the trunk.

Based on published studies, the intracerebral route may be more appropriate in some cases. For example, lymphoblastoid cells have been shown to proliferate best when inoculated by the intracerebral route.

7. **Observation period**
All animals are examined weekly by observation and palpation for a minimum of 16 weeks (*i.e.*, 4 months) for evidence of nodule formation at the site of injection when the route of inoculation is IM or SC. Examinations need not be more frequent than twice a week for the first 3 to 6 weeks, and once a week thereafter.

In some countries, the observation period is 4 to 7 months, depending on the level of concern associated with the specific cell substrate in the context of the product being developed. Whether a longer observation period is needed should be agreed with the NRA/NCL. Also see 8 below.

8. **Assessment of the inoculation site over time**
If a nodule appears, it is measured in two perpendicular dimensions, the measurements being recorded weekly to determine whether the nodule grows

progressively, remains stable, or decreases in size over time. Animals bearing nodules that appear to be regressing should not be sacrificed until the end of the observation period. Cell lines that produce nodules that fail to grow progressively are not considered to be tumourigenic.

If a nodule that fails to grow progressively, but persists during the observation period and it retains the histopathological morphology of a neoplasm, this should be discussed with the NRA/NCL to determine whether additional testing will be required. Such testing could include extending the observation period or switching to a newborn nude mouse, ATS-treated newborn rat, or other *in vivo* model to assess the tumourigenicity of the cell substrate.

If the cells that are injected fail to form tumors or to persist during the 4-month observation period, it might be necessary to extend the observation period or switch to a newborn nude mouse, ATS-treated newborn rat, or other *in vivo* model to assess the tumourigenicity of the cell substrate. This will depend on the level of concern associated with the specific cell substrate in the context of the product being developed. Whether such additional testing is needed should be agreed with the NRA/NCL.

9. **Final Assessment of the inoculation site and other sites**
   At the end of the observation period, or at an earlier time if required due to the death of an animal or other justifiable circumstances, all animals, including the reference group(s), are sacrificed and examined for gross and microscopic evidence of the proliferation of inoculated cells at the site of injection and in other sites such as the heart, lungs, liver, spleen, kidneys, brain, and regional lymph nodes since some CCLs may give rise to tumors at distant sites without evidence of tumor at the injection site. The tissues are fixed in 10% formol saline and sections are stained with hematoxylin and eosin for histological examination to determine if there is evidence of tumor formation and metastases by the inoculated cells.

10. **Assessment of metastases (if any)**
    Any metastatic lesions are examined further to establish their relationship to the primary tumor. If what appears to be a metastasis to a distant site differs histopathologically from the primary tumor, consideration should be given to the possibility that the tumor either developed spontaneously, or that it was induced by one or more of the components of the cell substrate such as an oncogenic virus.

If the histopathology or genotype of any tumors that develop are inconsistent with the inoculated cell type, or are of a histopathological type that has not been recognized as occurring spontaneously in the test species, additional tests should be undertaken to determine whether such tumors are actually spontaneous or are induced by elements within the cell substrate itself such as oncogenic viruses or oncogenic DNA sequences. In such cases, appropriate follow-up studies should be discussed and agreed with the NRA/NCL.

11. Interpretation of results

a. The test in nude mice is considered positive if at least 2 of 10 animals inoculated with the test article cells develop tumors that meet all of the following two criteria:

   i. Tumors appear at the site of inoculation or at a metastatic site
   ii. Histologic or genotypic examination reveals that the nature of the cells constituting the tumors is consistent with that of the inoculated cells

   In the past, chromosomal markers have been useful to demonstrate that the tumor cells are of the same species as that from which the inoculated cells were derived. However, the use of cytogenetics for this purpose has largely been replaced by genetic and antigenic markers.

b. If only one of 10 animals develops a tumor that meets the three criteria in 11.a, the cell line should be considered possibly tumourigenic and examined further. Such testing could include one or more of the following: a) repeating the test in an additional 10 nude mice, b) extending the observation period, c) increasing the size of the inoculum, or d) switching to the newborn nude mouse model, the ATS-treated newborn rat model, or other *in vivo* model. In such cases, appropriate follow-up investigations should be discussed and agreed with the NRA/NCL. For example, it may be appropriate to determine if the tumor is of nude mouse origin and whether there are any viral or inoculated cell DNA sequences present.

   Assessment of dose-response may provide additional information on the characteristics of the CCL. If such studies are undertaken, the design should be based on the *in vivo* titration of the inoculum in groups of 10 animals per dose level. For example, if 10 of 10 animals develop tumors with an inoculum of 10^7 cells, the titration could be done with 10^5, 10^3, and 10^1 cells in groups of 10 animals each.

Appendix 3

Oncogenicity Protocol for the Evaluation of Cellular DNA and Cell Lysates

When appropriate, and particularly for vaccines, cell DNA and cell lysates from tumorigenic cell substrates should be examined for oncogenicity in a test approved by the NRA/NCL.

In some countries, the following testing strategy is used.

1. **Type of test animals**
   Newborn (*i.e.*, <3 days old) nude mice, newborn hamsters, and newborn rats have been used to assess the oncogenic potential of cell lines. At this stage, it is not possible to draw definitive conclusions on the relative sensitivity of the three animal assays for oncogenicity, and testing is recommended in each of them. When data on the ability of these models to detect oncogenic activity are obtained, this recommendation may change.

2. **Point in the life history of the cells at which they should be tested**
   Cells from the MCB or WCB, propagated to the proposed *in vitro* cell age for production or beyond, should be examined for oncogenicity. Three extra population doublings ensure that the results of the oncogenicity test can be used in the assessment of overall safety of the product even under the assumption of a "worst case" situation and therefore provide a safety buffer.

3. **Use of controls**
   The purpose of the positive control is to assure that an individual test is valid by demonstrating that the animal model has the capacity to develop tumors from inoculated cell components (*i.e.*, a negative result is unlikely to be due to a problem with the *in vivo* test). While an appropriate positive control for cell-lysate oncogenicity assay is not clear, the recent description of an oncogene-expression plasmid for activated H-*ras* and c-*myc* has been shown to induce tumours in animals [1]. As the test with cell lysates is designed primarily to detect oncogenic viruses rather than oncogenes, the use of DNA as a positive control might not be suitable, both because of the nature of the assay and because DNA might not be stable in a cellular lysate.

   Whether a negative control arm, such as PBS, is included should be discussed with the NRA/NCL. An advantage of including a negative-control arm is that the frequency of tumour induction with lysates is expected to be low and may approximate the spontaneous tumour frequency in the indicator rodent, providing an important comparison to the test article arm.

4. **Number of test animals**
While the number of animals in a tumourigenicity test can be 10 per group, the number in an oncogenicity test should be larger due to the lower expected tumour incidence. The number per group should be discussed with the NRA/NCL.

5. **Inoculation of test material**
   a. **Cell lysate.** A lysate of the cells should be prepared by a method that avoids virus disruption while allowing maximum virus release and ensuring all cells are lysed (*e.g.*, three freeze/thaw cycles followed by low-speed centrifugation). Each animal should be inoculated subcutaneously above the scapula with a lysate obtained from 10^7 cells. Before inoculation, it should be determined that no viable cells are present, as development of tumours from cells would invalidate the test. The cell lysate is suspended in PBS and inoculated in a volume of 50-100 µL into newborn nude mice, newborn hamsters, and newborn rats [2]. If there is no evidence of a progressively growing tumour at the site of inoculation or at distant sites at the end of the observation period, the cell line may be considered not to possess oncogenic activity. If tumours are observed in this assay, the species of origin will need to be confirmed. The species of tumours that arise in a tumourigenicity assay will be that of the cell substrate, while the species of tumours that arise in an oncogenicity assay is that of the host (*e.g.*, rodent). If the cells were not lysed properly, it may be that the tumours that arose were from the species of the cell substrate.

   b. **DNA.** Total cellular DNA isolated from the cell substrate should be inoculated subcutaneously above the scapulae in PBS into newborn nude mice, newborn hamsters, and newborn rats. The amount of DNA inoculated should be ≥100 µg in 50 – 100 µL. Because of the concentrations necessary to achieve ≥100 µg of DNA, it might be necessary to shear the DNA; this can be done by sonication or by several passes in a needle and syringe. A positive-control plasmid with the test article DNA should be inoculated into a few mice to confirm that the cellular DNA is not inhibitory and that the animals are susceptible to tumour induction by DNA.

6. **Observation period**
Animals are examined weekly by observation and palpation for evidence of nodule formation at the site of injection. The observation period should be at least 4 months.

7. **Assessment of the inoculation site over time (progressive or regressive growth)**
   If one or more nodules appear, they are measured in two perpendicular dimensions, the measurements being recorded weekly to determine whether the nodule grows progressively, remains stable, or decreases in size over time. Animals bearing nodules that are progressing should be sacrificed when the nodule reaches a size of approximately 2 cm in diameter, unless a lower limit has been established by authorities for the humane treatment of animals.

8. **Final Assessment of the inoculation site**
   At the end of the observation period, all animals, including the reference group(s), are sacrificed and examined for gross and microscopic evidence of tumour formation at the site of injection and in other sites. Any tumour that is identified is divided into three equal parts: a) fixed in formalin for histopathology; b) used to establish a cell line, when possible; and c) frozen for subsequent molecular analysis.

9. **Evaluation of animals for metastases**
   Animals are examined for microscopic evidence of metastatic lesions in sites such as the liver, heart, lungs, spleen, and regional lymph nodes.

10. **Assessment of metastases (if any)**
    All tumours are examined to establish their relationship to the primary tumour at the site of inoculation. If what appears to be a metastatic tumour differs histopathologically from the primary tumour, it is necessary to consider the possibility that this tumour developed spontaneously. This may require further testing of the tumour itself. In such cases, appropriate follow-up studies should be discussed and agreed with the NRA/NCL.

11. **Interpretation of results**
    If tumours arise in the cell lysate or DNA assay, then these could be induced by an oncogenic virus or oncogenic DNA. Because of the implications for the use of a cell substrate that contains an oncogenic agent or an oncogenic activity for a biological, the NRA/NCL should be consulted to consider additional experiments to identify the oncogenic agent/activity and to determine the suitability of the use of the CCL.

12. **Applicability**
    Cell Banks: MCB or WCB taken beyond EOPC level/ECB
    Cell Types: CCL, SCL (recommended when tumourigenic cells are used in vaccine production)

References

1. Li Sheng-Fowler, Fang Cai, Haiqing Fu, Yong Zhu, Brian Orrison1, Gideon Foseh, Don G. Blair, Stephen H. Hughes, John M. Coffin, Andrew M. Lewis Jr and Keith Peden. *Int. J. Biol. Sci.* 2010; 6(2):151-162
2. Tatalick, L.M., Gerard, C.J., Takeya, R., Price, D.N., Thorne, B.A., Wyatt, L.M. and Anklesaria, P. (2005) Safety characterization of HeLa-based cell substrates used in the manufacture of a recombinant adenoassociated virus-HIV vaccine. *Vaccine.,* **23**: 2628-2638.

List of Abbreviations

ALS	anti-lymphocyte serum
ATG	anti-thymocyte globulin
ATS	anti-thymocyte serum
BHK-21	baby hamster kidney
bp	base pairs
BPL	beta-propiolactone
BSE	bovine spongiform encephalopathy
BVDV	bovine viral diarrhea virus
CCLs	continuous cell lines
CEF	chick embryo fibroblasts
CHO	Chinese hamster ovary
CJD	Creutzfeldt–Jakob disease
CPE	cytopathic effect
CTL	cytotoxic T-lymphocyte
CWD	chronic wasting disease
DCLs	diploid cell lines
EBV	Epstein-Barr virus
ECB	Extended Cell Bank
EFSA	European Food Safety Authority
EMA	European Medicines Agency
EOP	End of production
EOPC	End of production cells
EPIC-PCR	Exon Primed Intron Crossing-PCR
FFI	Fatal Familial Insomnia
GMPs	good manufacturing practices
GSS	Gerstmann-Sträussler-Scheinker syndrome
HA	haemagglutinating
HCP	hamster cheek pouch
HDCs	human diploid cells
HEK	human embryonic kidney

HERVs	human endogenous retroviruses
HLA	human leukocyte antigen
ICH	International Conference on Harmonization
IFA	immunofluorescence assays
IFN	interferon
IM	intramuscular
IPC	in process control
LCMV	lymphocytic choriomeningitis virus
MAbs	monoclonal antibodies
MCB	master cell bank
miRNA	microRNA
MOI	multiplicity of infection
NA	nucleic acid amplification techniques
NCLs	National Control Laboratories
NRAs	National Regulatory Authorities
OIE	World Organization for Animal Health
PCCs	primary cell cultures
PCR	polymerase chain reaction
PDL	population doubling level
PERT	product enhanced reverse transcriptase
PMCA	protein misfolding cyclic amplification
RCBs	Reference Cell Banks
rcDNA	residual cellular DNA
rDNA	recombinant DNA
RFLP	restriction fragment length polymorphism
RT	reverse transcriptase
s.c.	subcutaneous
SCs	stem cells
SCID	severe combined immunodeficiency
SCLs	stem cell lines
SG	Study Group

SIVB	Society for in Vitro Biology
SOP	standard operating procedure
SPF	specific-pathogen-free
STR	short tandem repeats
SV40	simian virus 40
TEM	transmission electron microscopy
TME	transmissible mink encephalopathy
TPD_{50}	tumor-producing dose at the 50% endpoint
TSEs	transmissible spongiform encephalopathies
TTV	transfusion-transmitted virus
VNTR	variable number of tandem repeats
WCB	working cell bank

www.ingramcontent.com/pod-product-compliance
Lightning Source LLC
Chambersburg PA
CBHW081111170526
45165CB00008B/2415

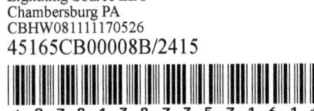